Cosmos and Universes

J. Balslev

Based on the doctoral thesis: The Structure and Composition of the Cosmos.

0.1 Preface

The book is based on the thesis: The Structure and Composition of the Cosmos, which consists of two theories, namely: the Quantum Ether Theory (QET) and the Euclidean Cosmos Theory (ECT), where the QET forms the foundation for the ECT that deduces the composition of the Cosmos. Taken together, the two theories give a description of the Cosmos and universes, that fits with the observations of our universe and makes it possible to understand how it all works.

The first half of the book, which deals with the QET, finds from the properties of space and time, that the space is Euclidean and that the relativistic laws are a result of a movement relative to the final propagation velocity of the forces in the zero-point field (ZPF). The second half of the book, entitled the ECT, derives the distribution of matter and energy in the infinite Euclidean space by logical deduction, and establishes that matter and energy accumulates into closed universes and barren objects, all of which are in a kind of equilibrium.

If the universes are not to end up as barren objects, they must be of sufficient size to create regenerative processes, such as AGNs, which in turn generate the cosmic microwave background (CMB) and the stellar nebulae from which the stars are born. As a result of the incessant regenerative processes, the density of the universes is very sparse, which also can be seen from the density of our own Universe. From this density, it is possible to make an estimation of the size of the closed universes, which is practically the same for all the universes which have reached their maximum size, because, when a universe becomes too large, it will not be able to hold on to any additional matter or energy.

The QET reintroduces the ether as a consequence of the Quantum Field Theory plus WMAP's and S. Marinov's measurements of the absolute velocity relative to the zero-point field (ZPF). Since the electromagnetic and gravitational forces propagate in the ZPF at the final speed of light, relativistic phenomena arise when bodies that are bound together by one of these forces have a velocity relative to the field and thereby relative to the propagation velocity of the forces that bind them together. In this way relativity and related phenomena can be deduced from classical physics such as Coulomb's Law and Newton's law of universal gravitation. This applies to the length contraction, the relativistic mass and the relativistic gravitational force, and from these laws it is proven, that the time is absolute and universal, so the space is Euclidean.

The ECT finds that according to the law of conservation of energy, the amount of energy is constant, and thereby final; and since the space is Euclidean, the Cosmos has always existed and consists of an infinite flat space where the total amount of matter and energy is collected. Because of gravity, matter and energy will, as time passes by, accumulate into a final number of barren objects, black holes and closed universes, which are all in a kind of equilibrium. However, even the closed universes will ultimately end up as barren objects, if the black holes were not able to distribute their matter and energy through regenerative processes; where a regenerative process is defined as a process that transforms heavy elements into lighter ones; and since the cosmological redshift mainly consists of a plasma redshift, which occurs as a result of the photon's transfer of energy to the plasma, the universes are seen to be static.

In each of the closed universes, the influence of gravity entails that most of the matter and energy ends up as galaxies, which mainly consist of black holes. This is due to the fact that the Cosmos has always existed and it only requires about 3 solar masses to create a black hole. As the energy is constant, there is a life cycle of energy in each of the universes, where the stars and the black holes at the center of the galaxies create the largest regenerative processes when the concentration of matter gets sufficiently high. The regenerative processes deliver energy to the life cycle of matter and radiation in the universe, where the new energy often ends up as nebulae or as the CMB. The gas nebulae are the first step on the road of stars, giants, white dwarfs, supernovae, neutron stars and black holes, where the energy once again ends up at the center of the galaxy. As the CMB reflects the regenerative processes, it reflects in this way the structure of the universes with their great walls and large voids, which are created by the law of least effort. Finally, the theory calculates the size of the energy production from the black holes at the center of the galaxies, and explains how matter and energy escape the black holes.

Contents

Chapter 1

Introduction

As the fundamental forces propagate in the zero-point field (ZPF) as virtual particles, the forces have a constant velocity relative to the field, so when a solid body moves in relation to the field, it also moves in relation to the forces that hold the body together. This applies to both the electromagnetic and the gravitational forces. As a movement relative to the ZPF creates a force on the body, we need to take a closer look at relativity, since it is essential for our understanding of space and time, and thereby the distribution of matter and energy in connection with "The Structure and Composition of the Cosmos". In this way, we end up with two theories: the Quantum Ether Theory (QET) and the Euclidean Cosmos Theory (ECT).

Based on the existence of the ZPF, the space can be proven to be Euclidean, and since the amount of energy is constant, according to the law of conservation of energy, the Cosmos has always existed. As the space is Euclidean it consists of an infinite flat space where the total amount of forces, matter and energy are collected. Because of gravity, the matter and energy will then, as the time passes by, accumulate into a finite number of barren objects, black holes, and closed universes, which are all in a kind of equilibrium. However, even the closed universes would ultimately end up as barren objects, if the black holes at the center of the galaxies were not able to distribute their energy through so-called regenerative processes, where a "regenerative process" is defined as a process that transforms heavy elements into lighter ones.

The energy from the regenerative processes primarily consists of protons, alpha-particles, and electromagnetic radiation, which end up as gas nebulae and cosmic background radiation, where the gas nebulae are the first step in an energy-cycle that later ends up as a black hole with a neutron star at the center. The cosmic background radiation in each of the universes mirrors the activity (and thereby the structure) of the universes, and as the measured redshift of the cosmic microwave background (CMB) is not the result of an expansion of a universe, but rather the result of a plasma redshift that occurs as a consequence of the photon's transfer of energy to the plasma (which for instance can be seen from the heating of the corona of our Sun), all the universes are static. Since the universes are static, and only can hold on to the energy if they are closed, it is possible to give an estimated calculation of the size of the universes.

1.1 The Quantum Ether Theory (QET)

Until the year 1905 when Einstein presented his Special Theory of Relativity, [1] it was among other H. A. Lorentz's theory of space and time, [2] which was predominant. At that time, it was generally believed that there existed a luminiferous ether, which was the medium for the propagation of light. The luminiferous ether, or just the ether, has now by virtue of the Quantum Field Theory experienced a renaissance, in which it is described as the energy of the system's ground state, and the quantum state with the lowest possible energy is called the zero-point energy. In connection with the QET, the vacuum state is designated as the zero-point field (ZPF), which is the ground state devoid of any wave-particles.

The experimental basis for the QET already exists. The existence of irreducible electromagnetic vacuum fluctuations, at absolute zero temperature, has been experimentally verified by the Casimir effect, [3] and, according to Maxwell's equations, [4] the electromagnetic field propagates with the constant speed, $c_0 = (\varepsilon_0 \mu_0)^{-1/2}$, in free space, where ($\varepsilon_0$) is the vacuum permittivity and (μ_0) is the vacuum permeability. Since this speed is very close to the speed of light, Maxwell wrote: "We can scarcely avoid the conclusion that light consists of the transverse undulations of the same medium which is the cause of electric and magnetic phenomena." Furthermore, since the vacuum permittivity, which also appears in Coulomb's law, is the capability of the vacuum to permit electric field lines, and the vacuum permeability is the ability of an electric current to generate a magnetic field in vacuum, the speed of light is a product of the properties of the ZPF.

Another experiment, which only can be explained by the existence of the ZPF, was performed by Stefan Marinov, who measured the difference of the speed of light in opposite directions. In so doing, he found that the absolute velocity of the experimental setup relative to the ZPF was equal to 362 ± 40 km/s (ref. [5], p. 36). The principle behind the experiment is shown in Figure 1.1. The measurement was later confirmed by NASA's space mission, entitled WMAP (Wilkinson Microwave Anisotropy Probe). From the CMB data, it could be seen, that our Solar System is moving at a speed equal to 369 ± 0.9 km/s relative to the reference frame of the CMB (ref. [6], p. 231).

Furthermore, it can be seen, that the Michelson and Morley experiment (ch. 4.1.4) confirms the existence of the ZPF. Since both light and the electromagnetic field that transfers the electromagnetic force, according to Maxwell's equations move with the speed of light c_0, there seem to be two solutions to the Michelson and Morley experiment: 1) The speed of light, and accordingly the velocity of the electromagnetic field, must be constant in relation to any experimental setup, and thereby to any object, if it is to ex-

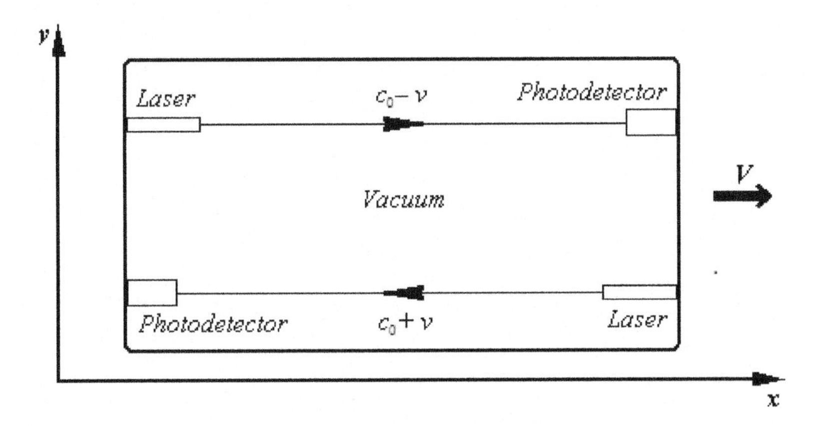

Figure 1.1: Measurement of the velocity relative to the zero-point field.

plain Michelson-Morley's experiment. 2) All bodies that move relative to the ZPF, and thereby relative to the propagation velocity of the electromagnetic field that holds the solid bodies together, must be subjected to a length contraction in the direction of motion.

If we choose the first option, such that light, and thereby the electromagnetic field, always has the same constant speed c_0 in relation to a solid body, the field will remain constant in relation to the body, so it will never be able to generate any length contractions. It is therefore only the second solution that works. It will later be shown, that a length contraction of the experimental setup is a solution to the Michelson-Morley experiment, and that the physical length contraction can be derived from Coulomb's law.

To show that the space is Euclidean, we consider a stationary system S and a moving system S' with the coordinates (x, y, z) and (x', y', z'), respectively, which coincide at the time 0. Let the x'-axis move along the x-axis with the relative speed v. According to Einstein's theory of relativity (ref. [7] p. 48) there then exists at that point of time the following relation: $t' = t\gamma(v) = t/\sqrt{1 - v^2/c_0^2}$ between the time t' in the moving system and the time t in the stationary system, and the relation: $x' = x\gamma(v) = x/\sqrt{1 - v^2/c_0^2}$ between the x'-coordinate in the moving system and the x-coordinate in the stationary system, where $\gamma(v) = 1/\sqrt{1 - v^2/c_0^2}$ is the Lorentz factor.

As the time is a measure of duration, it can be expressed by how long it takes to move a certain distance, i.e. $\Delta t = \Delta x/v$, where Δt is the time, Δx is the distance, and v is the speed. As the distance in the moving system $\Delta x'$ is altered by the factor $1/\sqrt{1 - v^2/c_0^2}$, the time $\Delta t'$ it takes to travel the distance $\Delta x'$ differs by exactly the same factor $1/\sqrt{1 - v^2/c_0^2}$, so the time will pass just as quickly in the moving system as in the stationary system:

$$v' = x'/t' = \frac{x/\sqrt{1 - v^2/c_0^2}}{t/\sqrt{1 - v^2/c_0^2}} = x/t = v.$$

In other words, if the speed v is the same in the two systems, so $v = v'$, it takes of course less time to move the shorter distance, so the time is exactly the same in the two systems. If it was not the case, we would have a situation, where the time in the x'-direction would be different from the time in the y'- and z'-direction, since the y' and z'-axes have not moved along the y- and z-axes. Consequently, we find that if we choose to look at space and time as a combined space-time, the time axis is just as linear as the three coordinate axes. This means that the time is universal, and since it is the time axis that creates the curvature of space, the space must be Euclidean, which is why gravity cannot be explained by the curvature of space-time.

To explain the structure of matter and the relativistic and gravitational observations, the QET is based on classical physics including the Quantum Theory. From the mass-energy relation (which is derived on the basis of classical physics in chapter 2.4.1) we have that:

$$E = mc_0^2,$$

where E is the energy, m is the mass, and c_0 is the speed of light, and from Max Planck's radiation law we have that:

$$E = h \cdot f,$$ [8]

where h is Planck's universal constant and f is the frequency of the emitted radiation, and finally from Maxwell's equations for the speed of light in vacuum:

$$c_0^2 = 1/(\varepsilon_0 \mu_0),$$ [4]

where ε_0 is the vacuum permittivity (also called the electric constant) and μ_0 is the vacuum permeability (also called the magnetic constant), we find that:

$$m = E/c_0^2 = h \cdot f/c_0^2 = \varepsilon_0 \mu_0 h \cdot f.$$

Since ε_0, μ_0, and h all are universal constants, it is seen, that the mass and inertia of a free wave-particle only depend on the frequency, which is true for both electromagnetic waves and solid bodies, and that matter is of electromagnetic nature. This wave-particle duality has been verified through a series of experiments, which includes both the electromagnetic radiation, neutrons, atoms and even molecules. [9] It means that the ZPF contains the prerequisites for creating all the chemical compounds within itself - including life.

Newton's law of universal gravitation states that two point masses m_1 and m_2 attracts each other with a force equal to $F_G = Gm_1m_2/r^2$, [10] where G is the universal gravitational constant, and r is the distance between the particles. If we make use of the expression found for the mass, the masses become equal to: $m_1 = \varepsilon_0\mu_0 h \cdot f_1$ and $m_2 = \varepsilon_0\mu_0 h \cdot f_2$, so the gravitational force equals:

$$F_G = G(\varepsilon_0\mu_0 h)^2 \cdot f_1 \cdot f_2/r^2.$$

From this expression for the gravitational force it can be seen that the force is of an electromagnetic nature, which is why the force propagates with the speed of light relative to the ZPF. Furthermore, it is seen that the gravitational force between two wave-particles only depends on the product of their frequencies and the square of the distance between them. For wave-particles it applies in general, that the higher the frequency becomes, the larger the mass becomes, i.e. the inertia and the gravitational force; where the inertia is the resistance against a change of the velocity of a wave-particle with the frequency f, and the gravitational force is the attraction towards other particles of a wave-particle with the frequency f. As the gravitational force from, and the inertia of, a wave-particle, both grow with the frequency of the particle, all wave-particles will receive the same acceleration in a gravitational field.

According to the QET, relativistic physical phenomena, such as the length contraction and the relativistic mass, arise, when solid bodies have a velocity relative to the ZPF and thereby relative to the propagation velocity of the electromagnetic forces that hold the bodies together. Since the gravitational field, just like the electromagnetic field, propagates with the final speed of light, c_0, in the ZPF, it is able to explain the observed relativistic phenomena in the Euclidean space, such as, for example, the peculiar motion of Mercury. With regard to the time, it cannot be a function of a physical length contraction. However, the clocks may be wrong as a result of a length contraction or a change of a gravitational force. The failure of the clocks depends on the speed and orientation of the clocks relative to their direction of motion and relative to the external forces.

Just as the length contraction is a result of a movement relative to the propagation velocity of the electromagnetic forces in the ZPF, it applies in general, that objects - which are bound together by forces that decrease with the square of the distance and have a constant propagation speed c_0 in relation to the ZPF - will be exposed to a relativistic contraction equal to $r(1 - v^2/c_0^2)^{1/2}$, when they have the rest length r and the speed v relative to the ZPF. Since the gravitational field also decreases with the square of the distance and propagates with the constant velocity c_0 relative to the ZPF, it also generates a relativistic contraction between two masses.

The relation between mass and frequency, where higher frequencies lead to greater masses of the wave-particles, is the basis for the distribution of matter and energy in the Euclidean space. If, in some way a wave-particle in free space arises, it will, regardless of the nature of the particle, gravitate towards the most forceful center of mass seen from the particle, so in a flat Euclidean universe matter and energy will gather in ever-greater collections, except for the places where the pressure gets so high that a regenerative process arises.

Altogether, the QET addresses the forces that arise when wave-particles move relative to the field and relative to each other, and thereby finds a natural explanation of black holes, gravitational redshifts, the deflection of matter and energy in a gravitational field, and the relativistic phenomena such as length contractions, time dilations, and relativistic forces.

1.2 The Euclidean Cosmos Theory (ECT)

The ECT looks at the distribution of matter and energy in the Euclidean space and is founded on classical physics and the following assumptions:

- The space is Euclidean.
- The law of conservation of energy holds good. [11]
- No EM interactions move faster than the speed of light in vacuum. [12]
- Matter and energy are deflected in a gravitational field.
- The Cosmos has existed for an infinitely long time. (Stems from the conservation law and the Euclidean geometry of space.)
- We exist.

Since the space is Euclidean - i.e. completely flat - the Cosmos consists of an infinite flat space where all the matter and energy are collected. When we look at the distribution of matter and energy, we assume that the Cosmos has existed for infinite length of time, that the total sum of matter and energy is constant, that no EM interactions move faster than the speed of light in vacuum, that matter and energy are deflected in a gravitational field, and that the matter and energy are quantized - and, therefore, cannot end up as a singularity. The gravitational force will then produce a distribution in the infinite flat space, where matter and energy gather in ever-larger and denser structures, until a state of equilibrium arises in the Euclidean space.

The larger and denser structures will, as a result of gravity, accumulate into black holes and closed universes, while the smaller structures, which are not able to hold on to the energy, will end up as barren objects when their energy is depleted. However, since the Quantum Theory does not permit singularities, even the closed universes will, when the energy is consumed, end up as giant black holes. But, since we exist, there must be a way in which a black hole can be converted into pure energy. That is to say, a black hole must be able to explode in a sort of regenerative process.

The larger a black hole becomes, the more of the surrounding matter it will be able to attract. Since a black hole from the outset has a mass, that is so big, that not even light can escape, all the energy that is fed into the black hole will stay there under normal conditions, so the black hole becomes bigger, denser and hotter. At some point the temperature and pressure may become so high, that a neutron star is formed at the interior of the black hole, if it was not there from the start. When the neutron star as a result of a still growing external pressure reaches the Tolman-Oppenheimer-Volkoff limit, [13] the neutron star will generate an explosion.

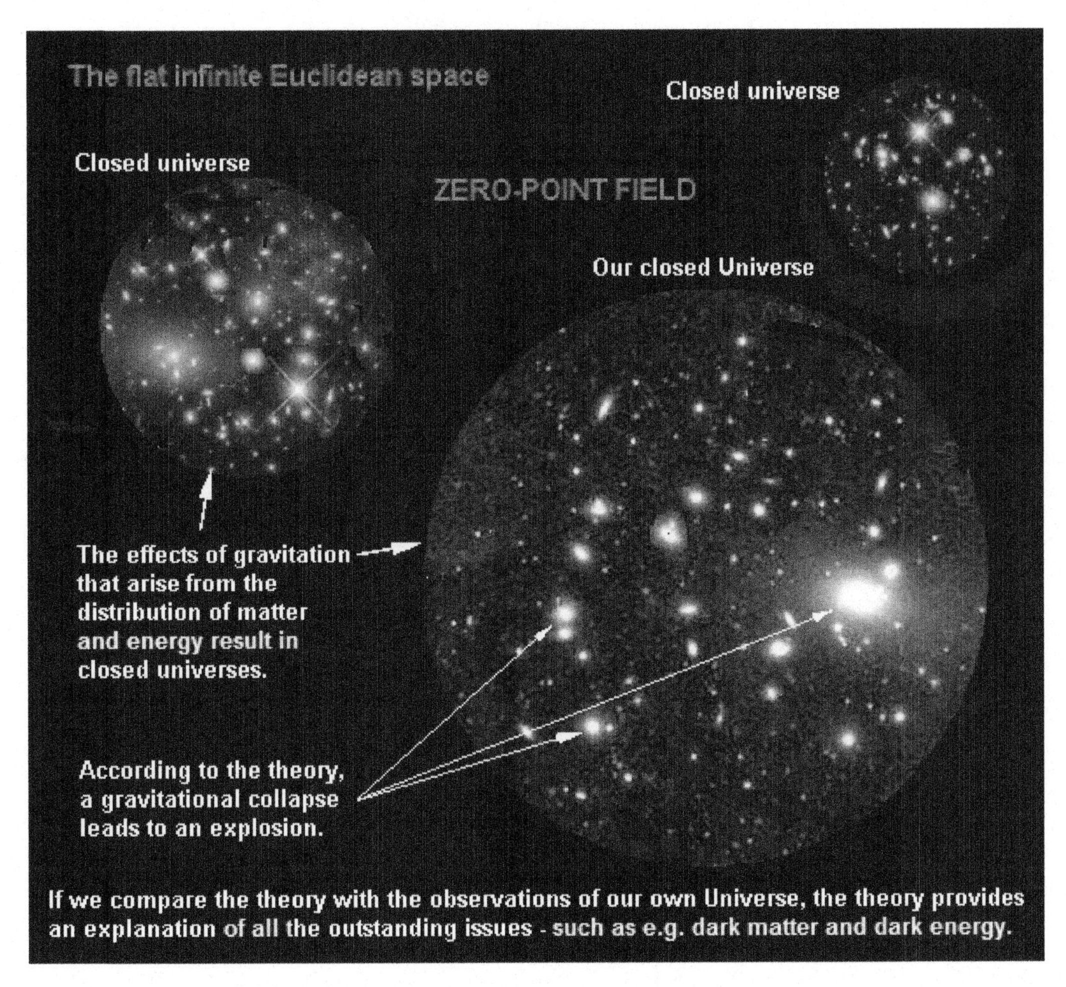

Figure 1.2: The Composition of the Cosmos.

The fission of neutrons into quarks and gluons releases large amounts of energy, where the quarks reach speeds around the speed of light. [14] This generates a pressure wave that, depending on the magnetic field and the rotational speed of the black hole with the neutron star at its center, will be able to reach the surface of the black hole near the direction of rotation, with velocities up to $\sqrt{3}c_0$ (ch. 3.4.7). Since the asymptotically free interacting quarks and gluons only can exist at extremely high pressures, the quark matter will, when the pressure falls, radioactively decay into stable particles through hadronization. [15] The produced hadrons and their decay products will, together with photons, be ejected from the black hole. [16] Such explosions, in an existing Euclidean closed universe, are among the most powerful regenerative processes..

The theory finds that the Cosmos consists of an infinite vacuum in which there are one or more closed universes. If there are more closed universes, they will all move away from each other, with velocities that for each of them are larger than the escape velocity from the overall system, or be in a relatively stable, dynamic equilibrium. It means that it may occur that universes collide. Moreover, in each of the closed universes regenerative processes will persistently occur. If a universe does not generate any regenerative processes, it will end up as a barren object. Since each of the universes has existed for a very long time, and since only around 3 solar masses are needed to create a black hole, [17] the universes will be teeming with black holes and burnt-out stars, which make up the framework of the galaxies that in turn make up the structure of the universes, with their super-clusters, filaments, great walls and giant voids.

When a black hole explodes in a regenerative process in an existing universe, the explosion generally takes place at the center of the galaxy, where the black holes with their neutron star at the center eject a hot, dense quark-gluon plasma. The resulting particles, such as photons, electrons, protons, and alpha particles end up as cosmic radiation, which provides the universe with new energy. Since the largest galaxies have the largest black holes and the most violent explosions, the galaxies are self regulating, in the way, that the largest galaxies expel most material, which sets an upper limit to the size of the galaxies.

In this way, the theory solves the flatness problem, [18] the horizon problem, and the smoothness problem, and explains the mass distribution in galaxies, and the network structure of the universes, as there is plenty of time to create the large structures by the law of least effort. Finally, the theory fits with the latest data from WMAP. [19]

Chapter 2

The Quantum Ether Theory

2.1 The Electromagnetic Field

The electromagnetic field is a component of the ZPF and is a fundamental force field that describes the electromagnetic interaction. It is a field produced by electrically charged objects, and affects the behavior of charged objects in the field.

The electromagnetic field can also be described as a sea of virtual particle pairs, which are predisposed by the surrounding electromagnetic forces, just as iron filings are affected by the electromagnetic field. This quantum picture of the electromagnetic field has proven to be very successful, giving rise to the Quantum Field Theory that describes the interaction of electromagnetic radiation with charged matter. [20]

In 1831 Michael Faraday made the observation, that time-varying magnetic fields could induce electric currents, and later in 1865 James Clerk Maxwell published his famous paper on "A Dynamical Theory of the Electromagnetic Field". [21]

2.1.1 Maxwell's equations

Maxwell's equations are a set of partial differential equations, which together with Coulomb's Law and Lorentz's force law form the foundation of classical electrodynamics. [22] Maxwell's equations describe how electric and magnetic fields are generated and altered by each other and by charges and currents. The equations are named after the Scottish physicist and mathematician James Clerk Maxwell, who published an early form of these equations between 1861 and 1862.

Maxwell's equations describe the large-scale behavior without having to consider the details of the atomic structure, but it requires in our case the use of parameters characterizing the electromagnetic properties of the ZPF.

Maxwell's equations: We start with Maxwell's equations in a vacuum, which consist of Gauss's law, Gauss's law for magnetism, Ampere's law and Faraday's induction law:

$$\nabla \cdot \vec{E} = \frac{\rho}{\varepsilon_0}, \; Gauss's\ law, \tag{2.1}$$

$$\nabla \cdot \vec{B} = 0, \; Gauss's\ law\ for\ magnetism, \tag{2.2}$$

$$\nabla \times \vec{B} = \mu_0 \varepsilon_0 \frac{\delta \vec{E}}{\delta t} + \mu_0 \vec{J}, \; Ampere's\ law, \tag{2.3}$$

$$\nabla \times \vec{E} = -\frac{\delta \vec{B}}{\delta t}, \; Faraday's\ induction\ law. \tag{2.4}$$

Here ∇ is the differential operator, $\vec{E}$ is the electric field, ρ is the charge density, ε_0 is the vacuum permittivity, $\vec{B}$ is the magnetic induction, μ_0 is the vacuum permeability, t is the time and $\vec{J}$ is the current density.

The four equations form a complete description of the production and interrelation of the electric and magnetic fields in a vacuum. The statements of the four equations are that: Equation 2.1 states that the electric field derives from electric charges, equation 2.2 states that there are no isolated magnetic poles, equation 2.3 states that changing electric fields and electric currents produces magnetic fields, and finally equation 2.4 states that changing magnetic fields produces electric fields.

Gauss's law relates the distribution of electric charge to the resulting electric field, and states that the measure of the electric field $\vec{E}$ through any closed surface is equal to $1/\varepsilon_0$ times the net electric charge ρ within that closed surface. The constant ε_0 is called the vacuum permittivity, where the value of the permittivity depends on the nature of the dielectric medium, which here is the ZPF.

The permittivity in the reference medium of the ZPF is equal to: $\varepsilon_0 = 8.8541 \times 10^{-12}$ F/m. [23] Since the permittivity is the ability of the ZPF to store electrical energy in an electric field, the ZPF cannot be empty, so the ZPF has properties which show that the space contains an ether. The vacuum permittivity is also described as the capacitance of the vacuum, as it is the capacitance which is present when creating an electric field in the ZPF.

The permeability in the ZPF is yet another proof of the existence of an ether, as the vacuum permeability $\mu_0 = 1.2566 \times 10^{-6}$ H/m can be described as a measure of the ability of the ZPF to support a magnetization within itself in response to an applied magnetic field.

2.1.2 Maxwell's equations in the Zero-Point Field

Since the charge density ρ is equal to zero in the ZPF, Gauss's law is reduced to:

$$\nabla \cdot \vec{E} = 0\,, \tag{2.5}$$

and since there are not any charges in the ZPF the current density $\vec{J}$, which is a measure of the density of the electric current, is also zero, this is why Ampere's law becomes:

$$\nabla \times \vec{B} = \mu_0 \varepsilon_0 \frac{\delta \vec{E}}{\delta t}\,. \tag{2.6}$$

So Maxwell's equations of the ZPF are reduced to:

$$\nabla \cdot \vec{E} = 0, \qquad \nabla \cdot \vec{B} = 0, \qquad \nabla \times \vec{B} = \mu_0 \varepsilon_0 \frac{\delta \vec{E}}{\delta t}, \qquad \nabla \times \vec{E} = -\frac{\delta \vec{B}}{\delta t}.$$

As the vacuum permittivity ε_0 is the ability of the ZPF to store electrical energy, the vacuum cannot be empty, and since the vacuum permeability μ_0 is the ability of the ZPF to support magnetic fields within itself, Maxwell interpreted it in the way that the electromagnetic waves do not arise out of nothing, but move like waves in the ZPF. The waves propagate, because a changing magnetic field generates a changing electric field according to Faraday's laws of induction. This electric field now creates a changing magnetic field according to Ampere's law. This continuous change between the electric and magnetic field allows the electromagnetic waves to propagate through the ZPF with the speed of light.

In the reference medium of the ZPF the constant ε_0 is the absolute permittivity of free space with the value: $\varepsilon_0 = 8.854187817 \times 10^{-12}$ F/m, [23] which entails that the ZPF is a dielectric. [24]

The vacuum permeability can be found experimentally, and even when the measurements take place in atmospheric air, the found "vacuum" permeability $\mu_0 = 1.297 \pm 0.140 \times 10^{-6}$ H/m lies within 97% of the actual vacuum permeability $\mu_0 = 1.2566 \times 10^{-6}$ H/m. [25]

2.1.3 The Electromagnetic Wave Function in the Zero-Point Field

Since the electromagnetic field is not under the influence of any charges or currents, we consider the equation (2.6):

$$\nabla \times \vec{B} = \mu_0 \varepsilon_0 \frac{\delta \vec{E}}{\delta t} .$$

A trivial solution to this equation is $\vec{B} = \vec{E} = 0$. However, beyond this solution Maxwell's equations also allow solutions of varying electric and magnetic fields, such as the wave equations for the electric and magnetic field, which can be written as:

$$\nabla^2 \vec{E} = \mu_0 \varepsilon_0 \frac{\delta^2 \vec{E}}{\delta t^2} \textit{ and } \nabla^2 \vec{B} = \mu_0 \varepsilon_0 \frac{\delta^2 \vec{B}}{\delta t^2} . \tag{2.7}$$

Since $\vec{D} = \varepsilon_0 \vec{E}$ and $\vec{B} = \mu_0 \vec{H}$, the wave equations (2.7) can also be written as:

$$\nabla^2 \vec{D} = \mu_0 \varepsilon_0 \frac{\delta^2 \vec{D}}{\delta t^2} \textit{ and } \nabla^2 \vec{H} = \mu_0 \varepsilon_0 \frac{\delta^2 \vec{H}}{\delta t^2} . \tag{2.8}$$

The general expression for a wave equation is:

$$\nabla^2 u = \frac{1}{c^2} \frac{\delta^2 u}{\delta t^2} , \tag{2.9}$$

where u is a scalar function, whose values could model the displacement of a wave and where the constant c is the propagation velocity of the wave. Since the product $\mu_0 \varepsilon_0$ has a unit equal to the inverse square of the velocity, and since the value of $(\mu_0 \varepsilon_0)^{-\frac{1}{2}}$ is almost equal to the speed of light as measured by Fizeau, [26] Maxwell showed in the early 1860s, that, according to the Theory of Electromagnetism he was working on, the electromagnetic waves propagate in empty space at a velocity that is almost equal to Fizeau's speed of light, which is why he proposed that light is in fact an electromagnetic wave. [27]

It can thus be seen, that there is the following relation between the vacuum permeability μ_0, the vacuum permittivity ε_0, and speed of light c_0 in vacuum:

$$\mu_0 \varepsilon_0 = 1/c_0^2 . \tag{2.10}$$

These electromagnetic waves do not arise out of nothing, but move like waves in the ZPF. The waves propagate, because a changing magnetic field generates a changing electric field according to Faraday's laws of induction. This electric field now creates a changing magnetic field according to Ampere's law. This continuous change between the electric and magnetic field allows the electromagnetic waves to propagate through the ZPF with the velocity c_0.

So, the vacuum permeability μ_0 can also be determined by using the relation $\mu_0 = 1/(c_0^2\varepsilon_0)$, whereby the vacuum permeability is given the value: $\mu_0 = 1.2566 \times 10^{-6}$ H/m in the SI system. And the electromagnetic waves can be described by a wave equation with amplitude Ψ, where:

$$\nabla^2\Psi(\vec{r},t) = \frac{1}{c_0^2}\frac{\delta^2}{\delta t^2}\Psi(\vec{r},t)\,. \tag{2.11}$$

A solution to this equation is a wave function of the form $\Psi(\vec{r},t) = Ae^{i(\vec{k}\cdot\vec{r}-\omega t)}$, in which:

$\Psi(\vec{r},t)$ is the wave function.
i is the imaginary unit.
k is the wave number ($k = 2\pi/\lambda = \omega/c$).
ω is the angular frequency, where $\omega = 2\pi \cdot f$, and f is the frequency.

It can be seen that the wave function is a solution to the wave equation, by inserting the wave function in the differential equation. If we choose to look at a solution along the x-axis, the right-hand side of the wave equation becomes equal to:

$$\frac{\delta^2}{\delta x^2}\Psi(x,t) = \frac{\delta}{\delta x}(\frac{\delta}{\delta x}Ae^{i(k\cdot x-\omega t)}) = \frac{\delta}{\delta x}(ikAe^{i(k\cdot x-\omega t)}) = -k^2Ae^{i(k\cdot x-\omega t)} = -k^2\Psi(x,t),$$

and the left-hand side is equal to:

$$\frac{1}{c^2}\frac{\delta^2}{\delta t^2}\Psi(x,t) = \frac{1}{c^2}\frac{\delta}{\delta t}(\frac{\delta}{\delta t}Ae^{i(k\cdot x-\omega t)}) = \frac{1}{c^2}\frac{\delta}{\delta t}(-i\omega Ae^{i(k\cdot x-\omega t)}) = -\frac{\omega^2}{c^2}\Psi(x,t).$$

Since the wave number $k = 2\pi/\lambda = \omega/c$, we get:

$$\frac{1}{c^2}\frac{\delta^2}{\delta t^2}\Psi(x,t) = -\frac{\omega^2}{c^2}\Psi(x,t) = -k^2\Psi(x,t),$$

where it only is the real part of the wave function that describes the field.

If we choose SI units, we find that light propagates in the ZPF with the velocity $c_0 = (\varepsilon_0\mu_0)^{-\frac{1}{2}}$, where both the vacuum permittivity ε_0 and the vacuum permeability μ_0 are constants associated with the electromagnetic ZPF. We can thus conclude, that the speed of light is constant in this medium. As $\varepsilon_0 = 8.854 \times 10^{-12}$ C^2/Nm^2 and $\mu_0 = 1.2566 \times 10^{-6}$ Ns^2/C^2, the speed of light becomes: $c_0 = 2.998 \times 10^8$ m/s. [24]

2.1.4 Coulomb's Law

Coulomb's law is a law of physics describing the electrostatic interaction between electrically charged particles. [28] The law has been heavily tested, and all observations have upheld the principle of the law. It was first published in 1785 by the French physicist Charles Augustin de Coulomb, and was essential to the development of the Theory of Electromagnetism. It can be used to derive Gauss's law, [29] and vice versa, and is in a way analogous to Isaac Newton's inverse-square law of universal gravitation. [10]

Coulomb's law states that the magnitude of the electrostatic force between two point charges is directly proportional to the scalar multiplication of the magnitudes of the charges and inversely proportional to the square of the distance between them. The force is along the straight line joining them, and if the two charges have the same sign, the electrostatic force between them is repulsive; whereas the force is attractive if the charges have different signs.

Coulomb's law specifically applies only when the charged bodies are much smaller than the distance separating them and can therefore be treated as point charges. In connection with quantum physics, Coulomb's law helps describe the forces that bind electrons to an atomic nucleus, that bind atoms together in molecules, and that hold solids and liquids together. The force is understood as arising from the electric field that surrounds the charges.

Coulomb's law states that the direct force F_C of point charge q_1 on point charge q_2, when the charges are separated by a distance r, is given by:

$$F_C = k_0 \cdot q_1 q_2 / r^2 \,, \tag{2.12}$$

where F_C is the force in Newtons, r is the distance in meters between q_1 and q_2, and $k_0 = 1/(4\pi\varepsilon_0)$ is Coulomb's constant, which is seen to include the vacuum permittivity ε_0.

2.2 The Quantum Field

The wave equation for electromagnetic fields corresponds to a wave equation for free particles. This equation is called the Schrödinger equation, and is the differential equation for free particles with mass m. The Schrödinger equation describes how the quantum state of a physical system changes with time. According to the standard interpretation of the Quantum Field Theory, the wave function is the most satisfactory description to be found of a physical system. The solutions to the Schrödinger equation describe the atomic and subatomic systems, atoms and electrons, as well as macroscopic systems.

The Schrödinger equation can be written as: [30]

$$\hat{H}\Psi(\vec{r},t) = i\hbar\frac{\delta}{\delta t}\Psi(\vec{r},t)\,, \tag{2.13}$$

where:

> $\hat{H}$ is the Hamilton operator, which is equal to the system's total energy E.
> $E = T + V$, where T is the kinetic energy and V is the potential energy.
> $\Psi(\vec{r},t)$ is the wave function.
> $\hbar$ is Dirac's constant, $\hbar = h/(2\pi)$, where h is Planck's constant.
> i is the imaginary unit.

By inserting the wave function $\Psi(\vec{r},t) = e^{i(\vec{k}\cdot\vec{r}-\omega t)}$, where the wave number $|\vec{k}| = k = 2\pi/\lambda = \omega/c$ and $\omega = 2\pi \cdot f$, it can be shown that the wave function satisfies the Schrödinger equation. If we just consider the function in the x-direction, the right-hand side becomes:

$$\hat{H}\Psi(x,t) = \hat{H}e^{i(k\cdot x-\omega t)},$$

and the left-hand side:

$$i\hbar\frac{\delta}{\delta t}\Psi(x,t) = i\hbar\frac{\delta}{\delta t}e^{i(k\cdot x-\omega t)} = \hbar\omega e^{i(k\cdot x-\omega t)}.$$

If the right- and left-hand side are to be equal, the solution must satisfy the following condition:

$$\hat{H}e^{i(k\cdot x-\omega t)} = \hbar\omega e^{i(k\cdot x-\omega t)},$$

so the Hamilton operator is equal to Dirac's constant times the angular frequency:

$$\hat{H} = \hbar\omega. \tag{2.14}$$

Since h-bar ($\hbar$) is equal to $h/(2\pi)$ and $\omega = 2\pi f$, we find that $\hbar\omega = hf = E$, that is:

$$E = fh. \tag{2.15}$$

This is called Max Planck's radiation law, which states that the total energy of a wave-particle E is equal to the frequency of the radiation f times Planck's constant h. It can be shown that the harmonic waves satisfy the wave function, as long as it applies that $c \cdot k = \omega$, or that the phase velocity c is equal to: [31]

$$c = \lambda \cdot f. \tag{2.16}$$

The Schrödinger equation (eq. 2.13) is a typical equation for the disturbance of a state of equilibrium and describes waves that propagate with the phase velocity c, which can be any velocity that satisfies the equation. Since the wave equation is linear in $\psi(x,t)$, it is satisfied by an arbitrary linear combination of functions, which are solutions to the equation. By superposition of harmonic waves with the proper relationship between ω and k, it is therefore possible to form new solutions to the wave equation such as wave packets, which make up the particles.

2.2.1 The Zero-Point Field (ZPF)

According to the Quantum Theory there is a restriction on the allowed evolution of quantum systems, which ensures that the sum of the probabilities of all possible outcomes of any event is always 1. This requirement, called unitarity, [32] is necessary for the consistency of the quantum system. Together with the law of conservation of energy, which states that energy can neither be created nor destroyed, it follows that the Hamiltonian cannot be zero. Since the second law of thermodynamics states that the total entropy of an isolated system always increases over time, the condition entails that the eigenstate of the Hamilton operator $\hat{H}$ (where $\hat{H}$ corresponds to the total energy of the system) is bound to a state of minimal energy called the ground state, which is not subject to change or decay to a lower energy state.

The ground state of the quantum field is also called the vacuum state or the ZPF. As a ground state the ZPF evidently has no kinetic energy that can provide it with a state of motion, which is why the ZPF does not take part in the movements of physical bodies. [33]

Although the ZPF generally possesses no physical particles, it may not always be empty. According to quantum physics, a quantum vacuum fluctuation is a temporary change in the amount of energy in a point in space, as explained by the uncertainty principle by Werner Heisenberg. According to this principle, energy and time can be related by the expression [34]

$$\Delta E \Delta t \approx \frac{h}{2\pi}. \tag{2.17}$$

Because of small fluctuations of the zero-point energy there may occur exchanges of tiny amounts of energy ΔE within short periods of time Δt. This allows for the creation of pairs of virtual particles. The effects of these particles are measurable such as, for example, in the effective charge of the electron that differs from its naked charge. [35]

From the energy sentence we know that it is not possible to create something out of nothing, so every supply of energy in one location must be followed by a reduction in another. If the field receives a supply of energy, the particles might even survive as real particles. If we look at the energy of the wave-particles, they will each have an energy equal to $E = hf$, [8] where Planck's constant is equal to $h = 6.62607015 \times 10^{-34}$ J·s. [36] From the expression $E = hf$ it can be seen that the energy E approaches zero when the frequency f approaches zero, so ultra-low-frequency electromagnetic waves below 3 Hz are among the waves with the lowest measured frequencies.

2.2.2 The Casimir Effect

Evidence of the existence of the ZPF is the Casimir effect that can be demonstrated experimentally. The Casimir effect is a very small attractive force that occurs between two uncharged conductive plates, when they are placed in the ZPF with a distance of a few micrometers. [3] The force arises because the standing waves in the intervening space between the parallel plates lead to a reduction of the vacuum pressure in the space between the plates, where the wavelengths of vacuum fluctuations are more limited than in the space around the plates.

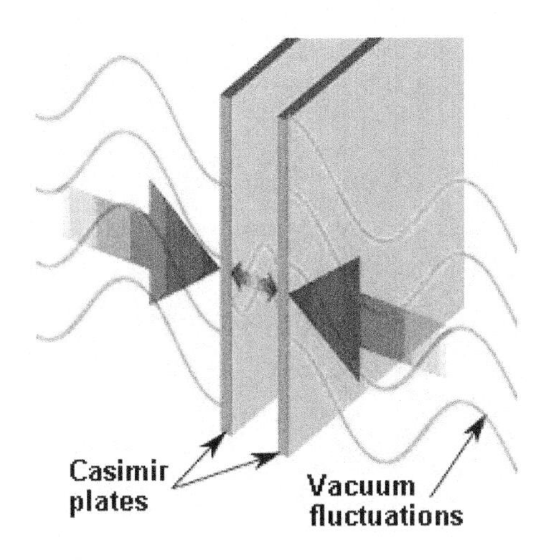

Figure 2.1: Casimir effect.

The wave-particles that make up our Universe, may, as the field is present everywhere, be regarded as excited states of the quantum field. In this way the ZPF serves as a medium in which the quantum waves and particles propagate and provides a background against which all velocities can be related. The particles can be energetic wave packets, such as photons and elementary particles, or force-carrying particles, such as virtual photons, gravitons and gluons.

Since the electromagnetic field, which holds the atoms together in a solid body, propagates with the speed of light in the ZPF, the velocity of the electromagnetic field will change in relation to the body, when the body moves relative to the ZPF. This changes the electromagnetic field surrounding the atoms. In order to maintain an equilibrium state, the atoms will then have to move relative to each other, until a state of equilibrium is obtained. This means that there is a physical explanation for length contractions.

2.2.3 The Speed of Light is Constant in the Zero-Point Field

As light consists of electromagnetic oscillations that propagate in the ZPF, it can be proven that the speed of light c_0' seen from a moving system S' is equal to the speed of light c_0 seen from the stationary system S. That is to say: *Light is always propagating in the ZPF with the constant speed c_0, which is independent of the state of motion of the observer and the emitting body.*

Assume that the systems S and S' coincide at the time $t_0 = 0$, and that the x'-axis in the system S' moves along the x-axis of S with the velocity v. Besides, let c_0' be equal to the speed of light, λ'_c be equal to the wavelength of light, and f_c' be equal to the frequency of light in the system S'.

Since the wavelength multiplied by the frequency equals the velocity of the wave, we have that $c_0' = \lambda'_c \cdot f_c'$ in the system S'.

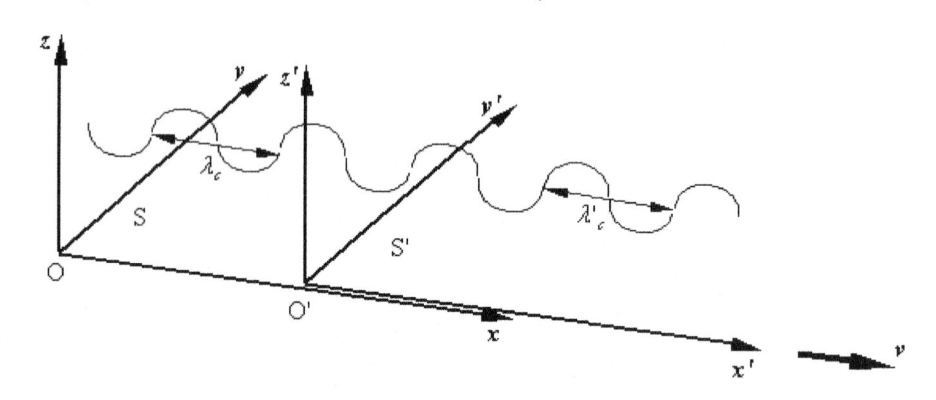

Figure 2.2: The speed of light in a reference system with velocity v.

During a period T the system S' has moved the distance $s = v \cdot T$ seen from S, where v is the velocity of S' in S. As $1/T = f_c$, the wavelength in S' is equal to:

$$\lambda'_c = \lambda_c - v \cdot T = \lambda_c(1 - v \cdot T/\lambda_c) = \lambda_c(1 - v/c_0)\,.$$

Since the wave motion seen from S does not change, the frequency in S' will be $1/(1-v/c_0)$ times larger, if the wavelength in S' becomes $(1 - v/c_0)$ times smaller, from which:

$$f_c' = f_c \frac{1}{1 - v/c_0}\,.$$

As the wavelength λ'_c multiplied by the frequency f'_c equals the speed of light c'_0 in the system S', we get:

$$c'_0 = \lambda'_c \cdot f'_c = \lambda_c(1 - v/c_0) f_c \frac{1}{1 - v/c_0} = \lambda_c f_c = c_0 \,. \tag{2.18}$$

This means that a measurement of the speed of light in an inertial frame of reference S', which has a velocity v relative to the ZPF, will register the velocity c_0.

Maxwell perceived the ether (the ZPF), as the medium in which light propagates. He regarded the electric field as a displacement of the ether from its equilibrium position, and meant that when such a displacement occurs, the ether itself will provide a restoring force that brings it back to its original resting position. This force is the cause of the magnetic field. [24]

If we compare the propagation of the electromagnetic field in the ZPF with something known, it may in a way be compared to the propagation of the sound in the air, because both the sound waves as well as the electromagnetic waves have a constant velocity in relation to a medium. But although objects can catch up with and break the sound barrier, they cannot catch up with the electromagnetic forces that bind them together. So objects that are bound together by electromagnetic forces cannot move faster than the speed of light.

2.2.4 Measurement of the Velocity relative to the Zero-Point Field

As light according to the QET propagates in the ZPF with the velocity c_0, the speed of light will be constant in relation to this medium. This property can therefore be used to measure the velocity v of an experimental setup relative to the ZPF, at the place on the Earth where the equipment is placed. Such an experiment has been performed by Stefan Marinov. [5]

The basic principle of Stefan Marinov's experiment is that laser light is divided and transmitted in small wave packages inside test equipment that moves in relation to the ZPF. Since the light propagates in the ZPF at the constant speed c_0, while the equipment is moving relative to the ZPF at the velocity v; the time it takes for the light to reach the photocell at the opposite end depends on the direction and velocity of the equipment in relation to the ZPF. So, if the experimental setup is placed along the direction of the Earth's motion relative to the ZPF, it will either take a shorter or longer time for the light to reach the opposite end. These time differences can be used to measure the velocity v of the equipment relative to the ZPF.

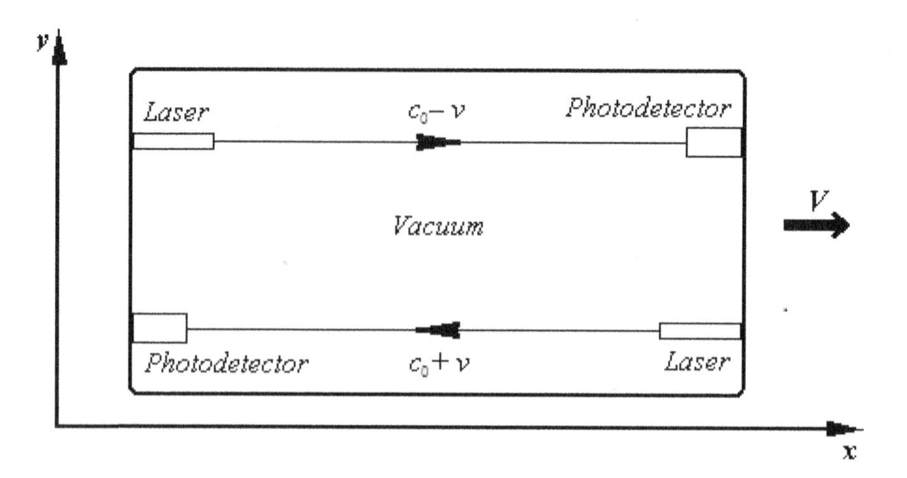

Figure 2.3: Measurement of the velocity relative to the ZPF.

The principle behind the performance of the measurement is shown in the drawing. The experimental setup consists of an evacuated chamber, so the drag velocity of any media, such as atmospheric air, can be ignored. The chamber is equipped with two lasers, one at each end, both of which emit packets of laser light; and in connection with each laser a photoelectric cell is placed at the opposite end of the chamber. When the chamber is at rest relative to the ZPF, the wave packets from the lasers will arrive simultaneously at each photocell.

However, if the vacuum chamber moves with a velocity v relative to the ZPF (as shown in Fig. 2.3) the laser light will get two different transmission speeds in relation to the vacuum chamber, since the light has a constant speed relative to the ZPF, which does not take part in the movement.

When a wave packet from a laser moves in the same direction as the chamber, the light will have to travel a longer distance before it reaches the photocell than when the light moves in the opposite direction, which applies although the chamber is exposed to a length contraction by the factor $\sqrt{1 - v^2/c_0^2}$. This feature can be used to determine the velocity of the experimental setup compared to the ZPF, i.e. the absolute velocity of the experimental equipment.

If we with help of photocells measure the electric current I_1 that comes from the photodetector in the direction of motion, and the current I_2 that comes from the photodetector at the opposite end, the difference:

$$\Delta I = I_1 - I_2 \,,$$

is a measure of, how fast the experimental setup moves relative to the ZPF.

If the experimental setup is rotated 180° around a vertical axis, we should get the same result as before (but with an opposite sign). In this way, any systematic errors from the experimental setup can be eliminated, such as differences in the construction of the photoelectric cells, etc.

2.2.5 Stefan Marinov's Coupled Shutters Experiment

In the following I will give an abbreviated version of the principles of Marinov's measurements of the absolute velocity of the Earth relative to the ZPF. Marinov's "coupled shutters" experiment consists of a rotating axle driven by an electro-motor, which is located at the middle of the axle. The axle has two rotating perforated disks, called hole-plates, one at each end of the rotating axle, so the plates are situated near the ends of an evacuated chamber which can be seen from the photo of the coupled-shutters experiment.

Figure 2.4: S. Marinov's coupled shutters experiment.

Light from a laser is divided, so the beams from the laser can pass through the holes of the discs in mutually opposite directions. The two hole-plates are completely identical and have the same holes, called windows, along the rim of the disk. The size of the windows is so small that they are completely covered by the laser beams, and the hole-plates are mounted in such a way that when the laser light passes through the window of the opposite plate it falls on the left rim of the window (seen from the center of the hole-plate) in such a way that exactly half of the window is covered by the laser light, when the axle is at rest, and the experimental setup is adjusted, so the setup is perpendicular to the Earth's velocity relative to the ZPF.

We will now look at an experimental setup, which is placed somewhere on the Earth. Since we do not know the size and direction of the velocity of the experimental setup, beyond that the velocity, regardless of how the Earth turns, will be greater than or equal to zero compared to the ZPF, we will make an adjustment to the experimental setup at the instantaneous velocity of the equipment relative to the ZPF, before the measurements are started. So, the hole-plates are kept still, while both laser beams are adjusted so that they each cover exactly half of the "window" on the opposite hole plate. As the laser light will hit the hole plate at exactly the same spot, as long as the experimental setup does not change its velocity or direction relative to the ZPF, it will be possible to perform the test if the equipment constantly points in the same direction.

Now the hole-plates will be set in rotation, so the laser light from both lasers is cut into pieces. Despite the experimental equipment having a velocity relative to the ZPF, the laser light will traverse the same path in both directions as during the adjustment. However, since the wave packets from the two lasers have to cover two different paths, because of the velocity of the equipment, the "windows" have time to move a longer or shorter distance, depending on how long it takes the wave packets to reach the windows and how fast the discs rotate. Since one of the wave packets, let us say the left, has the velocity $c_0 - v$ relative to the window, the right wave packet will have the velocity $c_0 + v$ relative to the other window, where the wave packet with the velocity $c_0 + v$ will reach the window faster than normal, which thus fails to rotate so much, which is why it will generate more current than normal.

We first consider the situation where v is equal to zero. The hole-plates will, at a constant speed of rotation and the constant velocity of the laser light, c_0, move a certain distance, which results in a certain current I_{1c} from the left window and a certain current I_{2c} from the right window. Since the distance the light must cover, is the same for both laser beams, and since the windows rotate equally fast, the currents must be the same when the velocity v equals zero. That is to say $I_{1c} = I_{2c}$ when $v = 0$. We can thus find the current I_c, which is generated when the experimental setup is stationary relative to the ZPF, by rotating the whole experimental setup around the center, until $I_c = I_{1c} = I_{2c}$. At this position the experimental setup stands perpendicular to the velocity of the apparatus relative to the ZPF. If we then turn the experimental setup 90 degrees, the setup will move with the velocity v, relative to the ZPF.

If the velocity v, is negative seen from the left laser, the left window will have time to rotate further than at the velocity $v = 0$, whereby the current from the left window decreases with the size $\Delta I_{1v} \leq 0$. This means that the current from the left window equals $I_{1v} = I_c + \Delta I_{1v}$, where $\Delta I_{1v} \leq 0$ when the absolute velocity $v \geq 0$. When the velocity v

is positive, seen from the right laser, the right window will not have time to rotate so far, as at the velocity $v = 0$, whereby the current from the right window becomes larger than I_c with the size $\Delta I_{2v} \geq 0$. This means that the current from the right window equals $I_{2v} = I_c + \Delta I_{2v}$, where $\Delta I_{2v} \geq 0$ when $v \geq 0$. Overall, we find, for the left and right window, that:

$$I_{1v} = I_c + \Delta I_{1v}, \text{ where } \Delta I_{1v} \leq 0; \text{ and } I_{2v} = I_c + \Delta I_{2v}, \text{ where } \Delta I_{2v} \geq 0 \text{ when } v \geq 0.$$

If the velocity v, is positive seen from the left laser, the left window will not have time to rotate so far, as at the velocity $v = 0$, whereby the current from the left window becomes larger than I_c with the size $\Delta I_{1v} \geq 0$. This means that the current from the left window becomes equal to $I_{1v} = I_c + \Delta I_{1v}$ where $\Delta I_{1v} \geq 0$ when the absolute velocity $v \leq 0$. When the velocity v is negative, seen from the right laser, the right window will have time to rotate further than at the velocity $v = 0$, whereby the current from the right window decreases with the size $\Delta I_{2v} \leq 0$. This means that the current from the right window equals $I_{2v} = I_c + \Delta I_{2v}$, where $\Delta I_{2v} \leq 0$ when $v \leq 0$. Overall, we find, for the left and right window, that:

$$I_{1v} = I_c + \Delta I_{1v}, \text{ where } \Delta I_{1v} \geq 0; \text{ and } I_{2v} = I_c + \Delta I_{2v}, \text{ where } \Delta I_{2v} \leq 0 \text{ when } v \leq 0.$$

The difference between the two currents I_{1v} and I_{2v} is a measure of how fast the experimental setup moves in relation to the ZPF:

$$\Delta I_v = I_{1v} - I_{2v} = \Delta I_{1v} - \Delta I_{2v}, \text{ where } \Delta I_{1v} \leq 0 \text{ and } \Delta I_{2v} \geq 0 \text{ when } v \geq 0$$
$$\text{and where } \Delta I_{1v} \geq 0 \text{ and } \Delta I_{2v} \leq 0 \text{ when } v \leq 0.$$

These currents can be measured by photocells. With a proper design of the profile of the laser beam, the windows, and the photocells, it is possible to construct the experimental setup so the velocity is proportional to the incoming light, which in turn is proportional to the current. It means that:

$$v = k\frac{\Delta I_v}{I_c}c_0, \text{ where } k \text{ is a proportionality constant.}$$

The result of the measurements is that the equipment moves relative to the ZPF at a velocity of 362 ± 40 km/s. [5] It is crucial, that we here have further proof of the existence of the ZPF, and that the ZPF is the basis for the propagation of light.

The measurement has later been confirmed by the Wilkinson Microwave Anisotropy Probe (WMAP) that found that our local group of galaxies appear to be moving at the velocity 369 ± 0.9 km/s relative to the CMB. [6]

2.3 Definition of Length and Time

Since, we cannot experience a lapse of time unless we have at least two consecutive physical events, the only option we have, when we want to define a unit of time, is to define time based on the velocity of a physical event.

The definition we consider must be consistent in relation to the stationary system. Since we have shown that the speed of light c_0 is constant in the ZPF, and independent of the motion of the source, it will be advantageous to use the speed of light, as the basis for the definition of a unit of time or a unit of length, and since the definition of the absolute time can best be based on periodic oscillations at a given frequency in the stationary system, the speed of light defines the length.

The International Standard Organization (SI) defines a second as the duration of 9192631770 oscillations of the radiation, corresponding to the transition between two hyperfine levels, of the ground state, of cesium 133. [37] This definition refers to a cesium atom at rest in the stationary system at a temperature of zero Kelvin. If Δt_p is the length of an oscillation period, then $\Delta t_p = 1 / 9192631770$ sec. So a second becomes equal to:

$$\Delta t_0 = 9192631770 \cdot \Delta t_p = 1 \text{ second}. \tag{2.19}$$

Based on the definition of a second and the speed of light in vacuum c_0, which is equal to 299792458 m per second, a meter m_0 can now be defined as: the length light travels in vacuum during the time interval $\Delta t = 1/299792458$ second. [37] From which it follows that a meter is equal to:

$$m_0 = c_0 \cdot \Delta t = c_0 \cdot 1/299792458 \text{ m} = 1 \text{ m}. \tag{2.20}$$

2.3.1 The Length Contraction

According to the QET, forces and particles are an intrinsic part of the quantum field in which they propagate. Since the electromagnetic field that generates the forces that hold the atoms together propagates with the velocity $c_0 = (\varepsilon_0\mu_0)^{-\frac{1}{2}}$ (eq. 2.10) in the ZPF, a body that moves relative to the field will also be moving relative to the electromagnetic forces that hold the body together. The solid body will thus be subjected to a length contraction.

This means that a physical reason can be found for the existence of length contractions. Since length contractions also affect the clocks, the length contractions can also lead to "time dilations", depending on the speed and orientation of the individual clock.

If we ignore external forces, the physical property that has an influence on the length of solid bodies is the electromagnetic forces that hold the bodies together. We therefore choose to look at Coulomb's law, and assume that we have two oppositely charged particles, or more precisely two oppositely charged point charges q_1 and q_2. We assume further that the particles, which are held together by electromagnetic forces, are at rest in the ZPF.

According to Coulomb's law, the force between the two charges is equal to:

$$F_C = k_0 q_1 q_2 / r^2 \,, \tag{2.21}$$

where F_C is the force, r is the distance between q_1 and q_2, and $k_0 = 1/(4\pi\varepsilon_0)$ is a constant, which from the electric constant ε_0 can be seen to be related to the ZPF.

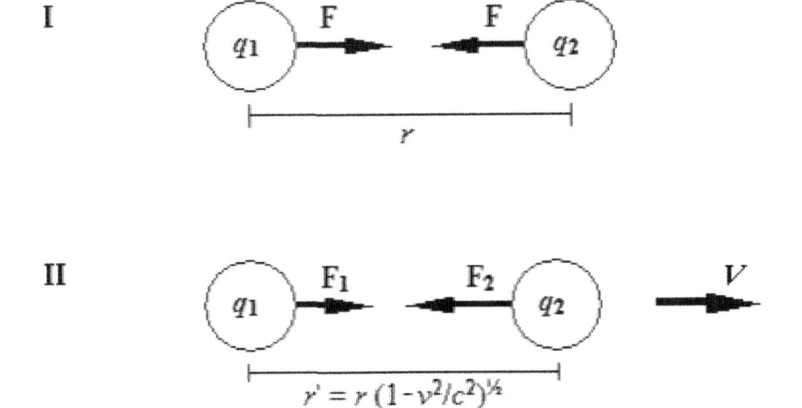

We know from earlier that the electromagnetic force propagates in the ZPF with the speed of light, c_0. This means that the distance r can be written as $c_0 \cdot t$, where t is the time it takes the force to move the distance r, i.e.:

$$r = c_0 t.$$

The two particles are then set in motion in the direction from q_1 to q_2, with the velocity v relative to the ZPF. Since both the force and the distance between the charges may have changed as a result of the movement relative to the field, we write:

$$F' = k_0 q_1 q_2 / r'^2 \,.$$

As the force propagates with the speed of light c_0, the distance, the force must travel from q_1 to q_2, becomes to the first order in vt: $r_1 = c_0 t + vt$, since the particle q_2 moves away from q_1. The distance, the force must travel from q_2 to q_1 becomes to the first order in vt: $r_2 = c_0 t - vt$, since the charge q_1 approaches q_2. We thus find:

$$F' = k_0 q_1 q_2/[r_1 r_2] = k_0 q_1 q_2/[(c_0 t + vt)(c_0 t - vt)] =$$
$$k_0 q_1 q_2/[c_0 t(1 + v/c_0)c_0 t(1 - v/c_0)]\,,$$

and since $c_0 t = r$,

$$F' = k_0 q_1 q_2/r'^2 = k_0 q_1 q_2/[r^2(1 - v^2/c_0^2)], \tag{2.22}$$

so:

$$r' = r\sqrt{1 - v^2/c_0^2}\,. \tag{2.23}$$

We have thus found the following relativistic expression for Coulomb's law:

$$F_C' = k_0 q_1 q_2/r'^2 = k_0 q_1 q_2/[r^2(1 - v^2/c_0^2)]. \tag{2.24}$$

So, the Lorentz contraction, or the relativistic length contraction that occurs when a body moves relative to the ZPF is equal to:

$$r' = r\sqrt{1 - v^2/c_0^2}. \tag{2.25}$$

The Lorentz contraction can be explained by the fact that the atoms and molecules that make up an object are held together by electromagnetic forces. Since, the electromagnetic field propagates in the ZPF with the constant speed of light, the velocity of the electromagnetic forces is altered in relation to the atoms of an object, when the object moves with a velocity v relative to the field. This exposes the object to a length contraction equal to $\sqrt{1 - v^2/c_0^2}$. We will therefore anchor the "stationary system" to the ZPF.

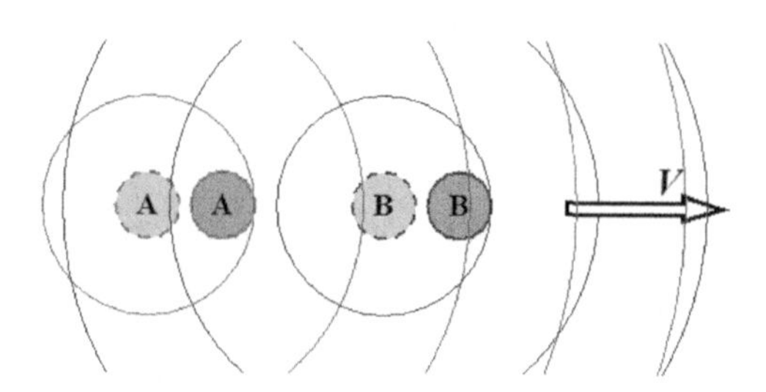

Figure 2.5: Two moving particles in the ZPF. The dotted lines symbolize a previous location, while the solid lines show the current position.

In general, it applies that objects - that are held together by forces that decrease with the square of the distance and have a final propagation velocity in the ZPF - will shrink when the objects move relative to the propagation velocity of the forces.

Let us consider two particles A and B that move with the common velocity v relative to the ZPF (Fig. 2.5). When the EM field from (B) reaches A, A has had time to approach (B), so A experiences a stronger force than if the objects had been static. Likewise, when the EM field from (A) reaches B, B has had time to distance itself from (A), so B experiences a weaker force than if the objects had been static. In other words, when A approaches the EM field from (B), the force on A will increase, and when B moves away from the EM field from (A), the force on B will decrease.

As the force is stronger, the closer the object is to the source of the EM field from the other object, the increase of the force on A, will be greater than the weakening of the force on B, so the total force becomes larger.

2.3.2 Visualization of an Object in Motion relative to the ZPF

As the complexity of the electrostatic forces between the individual particles of a solid body, which moves relative to the ZPF, is rather large, only a heuristic explanation of what takes place will be offered.

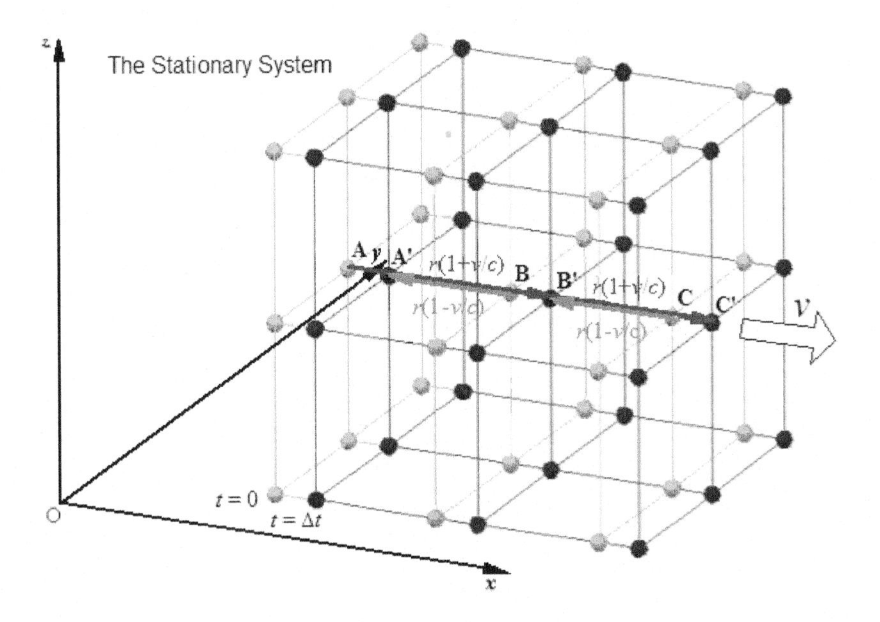

Figure 2.6: The electromagnetic forces inside a body with velocity v. Here, the light colored particles indicate an earlier position, whereas the black particles indicate the position a fraction of a second later.

Since the electromagnetic field propagates with the speed of light c_0 in the ZPF, the particles that constitute the body have time to move a certain distance before the electromagnetic forces from the neighboring particles arrive, which depends on the velocity of the body. Some of these forces are shown for particle **B**, in the middle of the solid body.

As shown in the drawing, the distance which the electromagnetic field must cover before it reaches particle **B**, is longest in the direction of motion (the blue arrows) and shortest in the opposite direction (the red arrows). Since the force becomes a little smaller when the distance increases, while the force becomes larger when the distance decreases, the total force will get larger.

The overall result is that the force from **A** to **B** becomes equal to the force from **B** to **C** and both forces will increase with a factor around $1/(1 - v^2/c_0^2)$, while the distances decrease with a factor about $(1 - v^2/c_0^2)^{\frac{1}{2}}$.

When the forces between the particles increase in the direction of motion, the particles will assume a new state of equilibrium, whereby the solid body will be exposed to a length contraction by a factor of around $(1 - v^2/c_0^2)^{\frac{1}{2}}$ in the direction of motion. Therefore, it is no wonder if the clocks are wrong.

2.3.3 "The Time Dilation"

Since the time t is a measure of duration, it can be expressed by how long it takes to move a certain distance r at the constant speed c, when r is at rest in the ZPF, i.e. $t = r/c$, where t is the time, r is the distance and c is the speed. It can be seen that the time is a linear function of the distance. However, if the distance r is subjected to a length contraction due to a velocity v relative to the ZPF, it will take less time to travel the shorter distance at the same speed.

Let us examine a system that moves with the velocity v relative to the ZPF. Due to the length contraction, the time t' it takes to travel the shorter distance $r' = r(1-v^2/c_0^2)^{\frac{1}{2}}$ in the moving system is not as long compared to the time t it takes to travel the distance r in the stationary system at the same speed c. This corresponds to what in the Theory of Relativity is called "the time dilation".

At the same speed c in the two systems, we find that:

$$c = r/t = r'/t' = r(1 - v^2/c_0^2)^{\frac{1}{2}}/t', \tag{2.26}$$

so:

$$1/t = (1 - v^2/c_0^2)^{\frac{1}{2}}/t', \tag{2.27}$$

from which we find "the time dilation":

$$t' = t\sqrt{1 - v^2/c_0^2}\,, \tag{2.28}$$

which in fact should be called the time contraction.

2.3.4 The Time is Absolute and Universal

Let Δx be a unit of length in the stationary system. The time Δt i takes the light to cover the distance Δx, is then:

$$\Delta t = \Delta x / c_0, \tag{2.29}$$

where c_0 is the speed of light in the ZPF.

Let us impart a velocity to the unit of length Δx along its axis, so it moves relative to the ZPF with a velocity v. According to the previously deduced Lorentz contraction (eq. 2.25), [2], [7] the length of Δx will then be equal to: $\Delta x' = \Delta x(1 - v^2/{c_0}^2)^{1/2}$. The time it takes the light to move the distance $\Delta x'$ then becomes:

$$\Delta t' = \Delta x'/c_0 = \Delta x(1 - v^2/{c_0}^2)^{1/2}/c_0 = \Delta t(1 - v^2/{c_0}^2)^{1/2},$$

from which we find that:

$$\Delta x'/\Delta t' = \Delta x(1 - v^2/{c_0}^2)^{1/2}/\Delta t(1 - v^2/{c_0}^2)^{1/2} = \Delta x/\Delta t = c_0.$$

That is to say, when the distance becomes shorter in the moving system, the time it takes to travel the shorter distance at the same speed c_0, becomes correspondingly shorter, so the time is reduced by exactly the same factor $(1 - v^2/{c_0}^2)^{\frac{1}{2}}$ as the length. So, the time will pass just as fast in the moving system as in the stationary system. This means that the time is the same in the two systems, so the time is absolute and universal, i.e., the time is the same everywhere in the Cosmos.

Let us look at an example in relation to the length contraction. Due to the rotational speed of the Earth, the circumference of the Earth at the equator will be subjected to a length contraction. So, if we travel a full circle around the equator, we can make the journey faster than if the Earth did not rotate. But, despite the fact that we arrive a little earlier, it does not mean we are exposing ourselves to a time dilation. So, if we call a friend at one of the Poles, which are not rotating, both parties will be at the same instant of time.

Since the presence of two or more consecutive events is the only way we can observe the flow of time, and since we wish to divide the constant flow of time into a series of equal periods, to make a precise measurement of the time, it will be appropriate to use an x-axis to specify the time intervals.

However, if we use a simple measuring rod that takes part in the movements of the Earth, to specify the time, the time intervals would fluctuate in accordance with the fluctuations of the length of the measuring rod, so the time intervals would be shorter or longer in accordance with the changes of the length intervals.

So, in connection with the definition of the time axis, we will make use of an x-axis with a scaling of equidistant divisions of the axis. There is then the following connection between the distance x, the speed of light c_0 and the time t:

$$x = c_0 \cdot t,$$

so the time may be expressed as:

$$t = x/c_0,$$

which is a linear function, since the speed of light c_0 is a constant. We could however have used any other constant velocity, if we had wished to.

2.3.5 The Space is Euclidean

If we consider the expressions for "the time dilation" $t' = t\sqrt{1 - v^2/c_0^2}$ and the length contraction $x' = x\sqrt{1 - v^2/c_0^2}$, it can be seen, that the time is shrinking by exactly the same factor as the length, so:

$$\frac{t'}{x'} = \frac{t\sqrt{1 - v^2/c_0^2}}{x\sqrt{1 - v^2/c_0^2}} = \frac{t}{x} = \frac{1}{c}.$$

This means that, at the constant speed c, it takes proportionally less time t' to travel the shorter distance x'. The time is therefore passing just as fast in a moving system as in a stationary system. So the time is absolute and universal. It would also be strange, if a physical length contraction had an impact on the flow of time.

If we choose to look at space and time as a combined space-time, the time axis will be just as linear as the three coordinate axes, so space-time will not curve. It means that the space is Euclidean. If we use a Cartesian coordinate system, we know from the previous chapter that the time can be expressed as a linear function of the distance x, so:

$$t = 1/c \cdot x. \tag{2.30}$$

Since the time axis is equal to a constant multiplied by a space axis, the time axis is just as linear as the three coordinate axes. So the combined space-time can best be described as an Euclidean space with three space axes and one time axis. The gravitational field cannot therefore be explained by a curvature of space-time. So, if the universe is closed, it will not be the space that bends, but the trajectories of matter and energy that are deflected in the gravitational field.

A closed universe is thus a universe from which neither matter nor energy can escape, but, as a result of the gravitational field, are drawn against the center of mass in the flat Euclidean space. It applies both to solid bodies as well as electromagnetic radiation, such as light, whose trajectories likewise are deflected in a gravitational field, [38] which is known from gravitational lensing. And since the space is perfectly flat, the gravity, which originates from the distribution of wave-particles, is transmitted by gravitons.

2.4 The Relativistic Mass

Let us consider a solid rectangular body, $x \cdot y \cdot z$, that is at rest relative to the ZPF and consists of atoms bound together by electromagnetic forces. Let the rest mass be equal to: $m_0 = \rho \cdot x \cdot y \cdot z$, where ρ is the density and x, y and z are the coordinates of the body in the respective directions, when the body is at rest in the ZPF.

We now impart the velocity v to the body - in the plus x direction. Because of the length contraction (eq. 2.25), the volume becomes equal to:

$$x' \cdot y \cdot z = x(1 - v^2/c_0^2)^{\frac{1}{2}} \cdot y \cdot z\,.$$

Since the number of elementary particles that the body consists of remains constant during a change of the velocity, and since we assume that the mass of the elementary particles remains constant during such a change, the total mass of the particles is assumed to be constant during a variation of the velocity. However, since the volume decreases due to the length contraction, the density becomes larger. Let us denote the new density with ρ'. It will then be equal to the rest mass m_0 divided by the new volume:

$$\rho' = m_0/[x' \cdot y \cdot z] = \rho \cdot x \cdot y \cdot z/[x(1-v^2/c_0^2)^{\frac{1}{2}} \cdot y \cdot z] = \rho/(1-v^2/c_0^2)^{\frac{1}{2}} = \rho(1-v^2/c_0^2)^{-\frac{1}{2}}\,.$$

If we instead of ρ insert the expression $m_0/(x \cdot y \cdot z)$, we find:

$$\rho' = m_0/[x' \cdot y \cdot z] = \rho(1 - v^2/c_0^2)^{-\frac{1}{2}} = m_0(1 - v^2/c_0^2)^{-\frac{1}{2}}/[x \cdot y \cdot z]\,,$$

so:

$$\rho' \cdot x \cdot y \cdot z = m_0(1 - v^2/c_0^2)^{-\frac{1}{2}}\,.$$

If we set $\rho' \cdot x \cdot y \cdot z$ equal to the mass m at the velocity v, we get:

$$m = \rho' \cdot x \cdot y \cdot z = m_0(1 - v^2/c_0^2)^{-\frac{1}{2}}\,,$$

from which we find Einstein's relativistic mass:

$$m = \frac{m_0}{\sqrt{1 - v^2/c_0^2}} = \gamma m_0\,, \tag{2.31}$$

where $\gamma = (1 - v^2/c_0^2)^{-\frac{1}{2}}$ is denoted as the Lorentz factor, and the mass m_0 is the rest mass of a body with the volume, $x \cdot y \cdot z$, where the body is at rest relative to ZPF.

If we impart the velocity v in the x-direction to a body that is at rest relative to the ZPF, it will shrink by the factor $(1 - v^2/c_0^2)^{\frac{1}{2}}$ in the x-direction due to the speed v, so the body during the movement gets the coordinates $x' \cdot y \cdot z$, and therefore the mass:

$$m = \rho' \cdot x' \cdot y \cdot z = \rho(1 - v^2/c_0^2)^{-\frac{1}{2}} \cdot x(1 - v^2/c_0^2)^{\frac{1}{2}} \cdot y \cdot z = \rho \cdot x \cdot y \cdot z = m_0\,.$$

That is to say, the mass remains the same, but the density becomes larger due to the reduced volume. This must be true for all matter that is bound together by forces that propagate with the speed of light c_0 relative to the ZPF. However, a complete description of the subject requires that we also look at the characteristics of the other fundamental forces and the behavior of the wave-particles, when they have a velocity relative to the stationary system.

2.4.1 Mass and Energy are Equivalent Quantities: $E = mc_0^2$

If a force F acts on a particle, the particle will have a change in momentum $p = mv$, which is greater the longer the force acts on the particle. Since the relativistic mass depends on the velocity v, we can express the force F as the change of the momentum per unit time: [39]

$$F = \frac{dp}{dt} = \frac{d(mv)}{dt}\,. \tag{2.32}$$

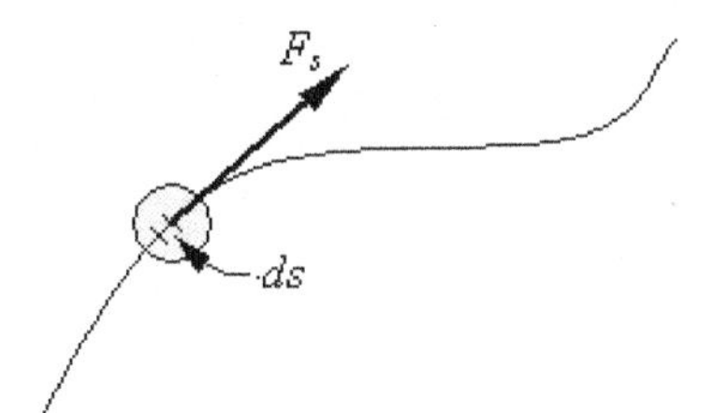

The work the force F_s performs on a particle that is not influenced by other external forces and which moves the distance ds along the trajectory of the particle, is equal to the increase of the kinetic energy of the particle, dE_k, where:

$$dE_k = F_s ds\,. \tag{2.33}$$

Using the chain rule on the force F_s, we find that:

$$F_s = \frac{dp}{dt} = \frac{d(mv)}{dt} = \frac{d(mv)}{ds} \cdot \frac{ds}{dt} = v \cdot \frac{d(mv)}{ds}\,,$$

from which:

$$dE_k = F_s ds = v \cdot d(mv) = v(vdm + mdv) = v^2 dm + mvdv\,.$$

If we insert the expression for the relativistic mass (eq. 2.31), where:

$$m = \gamma m_0 = \frac{m_0}{\sqrt{1 - v^2/c_0^2}}\,,$$

we find:

$$dE_k = v^2 \frac{dm}{dv} dv + mvdv = \frac{m_0 v^3/c_0^2}{(1 - v^2/c_0^2)^{\frac{3}{2}}} dv + \frac{m_0 v}{(1 - v^2/c_0^2)^{\frac{1}{2}}} dv = \frac{m_0 v dv}{(1 - v^2/c_0^2)^{\frac{3}{2}}}\,.$$

If we integrate from 0 to v and define, that $E_k = 0$ for $v = 0$, we find the following expression for the increase in the kinetic energy:

$$E_k = \int_0^v dE_k = \int_0^v \frac{m_0 v}{(1 - v^2/c_0^2)^{\frac{3}{2}}} dv = \frac{m_0 c_0^2}{(1 - v^2/c_0^2)^{\frac{1}{2}}} - m_0 c_0^2 = (m - m_0) c_0^2\,,$$

where $m_0 c_0^2$ is the rest energy of the particle.

The total energy of the particle (E) is then (when we ignore the potential energy) equal to the sum of its rest energy and the kinetic energy, E_k:

$$E = m_0 c_0^2 + E_k = m c_0^2\,.$$

So, the total energy of a particle is equal to its total mass m times the square of the speed of light, c_0^2:

$$E = mc_0^2. \tag{2.34}$$

That is to say, mass and energy are two sides of the same coin.

2.4.2 The Energy, Mass, and Frequency of Wave-Particles

The Compton effect shows that in connection with the scattering of very short wavelength electromagnetic radiation on electrons, the electromagnetic waves can be perceived as a flow of particles, each with the energy: [40]

$$E = hf, \tag{2.35}$$

where h is Planck's constant and f is the frequency, and with a momentum in the direction of propagation equal to: [33]

$$p = h/\lambda, \tag{2.36}$$

where h is Planck's constant and λ is the wavelength.

De Broglie assumed that the relations $E = hf$ and $p = h/\lambda$ are true for all wave-particles, e.g. photons as well as electrons, and suggested that material particles such as electrons, atoms and molecules could be seen as waves with a frequency $f = E/h$ and a wavelength $\lambda = h/p$. Since such waves can form interference patterns, we must assume that the superposition principle applies regardless of whether we deal with electromagnetic fields or material particles.

This means that in connection with energy and momentum of a particle, such as for instance an electron, there is a corresponding frequency and wavelength of a wave. Under the assumption that the waves propagate like classical waves, de Broglie showed that the waves only form standing waves for certain discrete frequencies, corresponding to discrete energy levels. One of the features of quantum mechanics is exactly this wave-particle duality, and it is clear that this duality is also present in the electromagnetic field.

Since f and λ in connection with electromagnetic radiation are connected through the relation $c_0 = f\lambda$ (eq. 2.18), we find that:

$$E = hf = p\lambda \cdot c_0/\lambda = p \cdot c_0,$$

and since the momentum p in connection with electromagnetic radiation with the velocity c_0 is equal to mc_0, we find that:

$$E = p \cdot c_0 = mc_0 \cdot c_0 = mc_0^2, \tag{2.37}$$

from which it follows that the total energy of electromagnetic radiation is equal to its mass multiplied by the square of the speed of light, just as for solid bodies -, i.e., the mass-energy equivalence is also true for electromagnetic radiation.

The total energy of a wave-particle must be equal to its rest energy relative to the ZPF plus the energy that is added to the particle, such as kinetic energy relative to the ZPF and potential energy according to its position relative to external forces, so the total energy of a wave-particle is equal to:

$$E = E_0 + E_k + U, \tag{2.38}$$

where E_0 is the rest energy, E_k is the kinetic energy and U is the potential energy.

According to the mass-energy equivalence (equation 2.34) and De Broglie's assumption that Planck's radiation law $E = hf$ are true for all wave-particles, we find that the energy of a wave-particle which is not subject to any external forces and at rest relative to the ZPF can be expressed as:

$$E_0 = m_0 c_0^2 = f_0 h.$$

The total energy of a wave-particle, when the kinetic and potential energy of the wave-particle is taken into account, is then equal to:

$$E = mc_0^2 = fh = E_0 + E_k + U = m_0 c_0^2 + E_k + U = f_0 h + E_k + U,$$

where E_k is the kinetic energy, and U is the potential energy. The mass and frequency of a wave-particle can then be expressed as:

$$m = m_0 + E_k/c_0^2 + U/c_0^2 \text{ and } f = f_0 + E_k/h + U/h.$$

The kinetic energy of a non-rotating wave-particle of mass m traveling at the speed v along a straight line in the ZPF is equal to the energy it possesses due to its motion. So the kinetic energy E_k is equal to the work done by a force F to accelerate the mass m to the velocity v.

$$E_k(v) = \int_0^v F dx = \int_0^v madx = \int_0^v m\frac{dv}{dt}\frac{dx}{dt}dt = \int_0^v mvdv = \frac{1}{2}mv^2.$$

The gravitational potential energy is the energy an object has because of its position in a gravitational field, and is equal to the work done against the force of gravity $F = -GmM/r^2$, where $G = 6.67 \times 10^{-11}$ m^3 kg^{-1} s^{-2}. Because of the inverse square nature of gravity, the force approaches zero at large distances from the gravitational field, which means that the increase in gravitational potential energy likewise approaches zero at infinity, so it makes sense to choose the gravitational potential energy equal to zero at infinity.

That is to say, it is reasonable to set the gravitational potential energy equal to zero in a region of the ZPF that is depleted of objects, whereby the potential becomes less than zero and thereby minus wherever there is a concentration of matter.

The potential energy of an object of mass m, which is located in a gravitational field, is then equal to the work done against the gravitational force F in bringing the mass in from infinity - where the potential energy is equal to zero - to its position in space equal to r. So the gravitational potential energy (U) equals:

$$U(r) = \int_\infty^r -Fdr = GMm\int_\infty^r 1/r^2 dr = -GMm[1/r]_\infty^r = -GMm/r.$$

The total energy of a wave-particle with the velocity v, which is located in a gravitational field that originates from the gravitational force $F = -GMm/r^2$, is then equal to:

$$E = mc_0^2 = fh = m_0c_0^2 + mv^2/2 - GMm/r = f_0h + mv^2/2 - GMm/r,$$

where $mv^2/2$ is the kinetic energy and $-GMm/r$ is the gravitational potential energy.

From the expression for the energy we find the mass of the wave-particle:

$$mc_0^2 = m_0c_0^2 + mv^2/2 - GMm/r \Leftrightarrow m = m_0 + \frac{mv^2}{2c_0^2} - \frac{GMm}{rc_0^2}$$

$$\Leftrightarrow m(1 - \frac{v^2}{2c_0^2} + \frac{GM}{rc_0^2}) = m_0,$$

where m_0 is the rest mass of the particle, v is its velocity relative to the ZPF and r is its distance from the center of gravity with mass M.

So the relativistic mass of a wave-particle, with the velocity v relative to the ZPF, that is located in a gravitational field at the distance r from the center of mass with the mass M, is:

$$m = m_0/[1 - v^2/(2c_0^2) + GM/(rc_0^2)]. \tag{2.39}$$

Since $m = fh/c_0^2$ and $m_0 = f_0h/c_0^2$ we find that the frequency of a wave-particle, with the velocity v relative to the ZPF, that is located in a gravitational field at the distance r from the center of mass with the mass M, is:

$$f = f_0/[1 - v^2/(2c_0^2) + GM/(rc_0^2)], \tag{2.40}$$

and, since $m = E/c_0^2$ and $m_0 = E_0/c_0^2$ we find that the energy of a wave-particle, with the velocity v relative to the ZPF, that is located in a gravitational field at the distance r from the center of mass with the mass M, is:

$$E = E_0/[1 - v^2/(2c_0^2) + GM/(rc_0^2)]. \tag{2.41}$$

It is noted that the equations for the mass, frequency, and energy of a wave-particle all follow the same pattern. Let us look at a situation where the speed of a wave-particle is increased. We then have to take relativity into consideration. If we ignore the gravitational potential energy, it can be seen from the expressions for the energy, frequency, and mass, that both the energy, frequency, and mass are increased by the factor $1/[1 - v^2/(2c_0^2)]$, when the speed v of the wave-particle is increased.

So both the energy, the frequency, and the mass of a wave-particle are dependent on the velocity. When the velocity v of a wave-particle approaches c_0, it is seen from the above, that both the mass, frequency, and energy are doubled. This dependency of the velocity corresponds to the relativistic change of composite solid bodies, which are bound together by forces that decrease with the square of the distance, which we previously have shown shrink with the Lorentz factor $\gamma^2 = 1/(1 - v^2/c_0^2)$.

In connection with heavy celestial objects, such as black holes and neutron stars, the gravitational potential energy has, together with the kinetic energy, a significant influence on both the mass, frequency and energy of the wave-particles, both inside, and in the vicinity of the heavy bodies.

2.4.3 Mass is a Consequence of the EM Oscillations in the ZPF

According to Maxwell's connection between the vacuum permittivity ε_0, the vacuum permeability μ_0, and the velocity c_0 of the electromagnetic waves, we have:

$$\varepsilon_0\mu_0 = 1/c_0^2\,,$$

and as $E = mc_0^2$ (eq. 2.37), we obtain the following expression for the mass:

$$m = \varepsilon_0\mu_0 E\,. \tag{2.42}$$

Furthermore, according to Max Planck's radiation law and de Broglie's description of material particles, the energy E of a material particle can be expressed as hf, where h is Planck's constant, and f is the frequency of the wave packet, such that:

$$E = h \cdot f\,.$$

The mass m of a wave-particle can then be expressed by:

$$m = \varepsilon_0\mu_0 h \cdot f\,. \tag{2.43}$$

From $m = \varepsilon_0\mu_0 hf$, it can be seen, that the mass is a consequence of the electromagnetic quantum energy of a wave-particle with the frequency f, where $\varepsilon_0 = 8.85 \times 10^{12}$ F/m, $\mu_0 = 4\pi \times 10^{-7}$ H/m and $h = 6.63 \times 10^{34}$ Js, all are constants related to the nature of the electromagnetic quantum field.

As the mass only depends on the frequency, f, it is obvious that long-waved electromagnetic wave-particles have such a small frequency that they are virtually massless, while gamma rays, electrons, and hadrons are quantum particles with a relatively high frequency and mass. The higher frequencies the wave-particles have, the greater gravitational attraction and inertia they acquire. This is independent of whether the wave-particles act as electromagnetic radiation or solid bodies.

2.4.4 Gravity expressed by the EM Oscillations in the ZPF

According to Newton's law of universal gravitation, the gravitational force is proportional to the product of the masses and inversely proportional to the square of the distance between them,

$$F_G = G \cdot m_1 m_2/r^2. \tag{2.44}$$

Here, F_G is the force between the masses m_1 and m_2, $G = 6.67408 \times 10^{-11}$ N m^2 kg^{-2} is the gravitational constant [41], and r is the distance between the centers of mass of the particles. From the following expression for the mass m of a wave-particle (equation 2.43):

$$m = \varepsilon_0 \mu_0 h \cdot f,$$

we can find an expression for the force between the particles based on a constant, the frequencies of the particles and the distance between them. As the masses can be expressed as $m_1 = \varepsilon_0 \mu_0 h \cdot f_1$ and $m_2 = \varepsilon_0 \mu_0 h \cdot f_2$, the gravitational force F_G can be written as:

$$F_G = G(\varepsilon_0 \mu_0 h)^2 \cdot f_1 \cdot f_2 / r^2. \tag{2.45}$$

Consequently, the force between the particles can be expressed by the gravitational constant G, the vacuum permittivity ε_0, the vacuum permeability μ_0, Planck's constant h, the frequency of the quantum waves f_1 and f_2, and the distance between the wave-particles. If it is necessary to find the total force between two solid bodies on the basis of the frequencies of the individual quantum particles, it will be necessary to make a summation over all the quantum particles of the bodies.

From Newton's law of universal gravitation, $F_G = GMm/r^2 = ma$, it can be seen, that the acceleration $a = GM/r^2$ of the mass m is independent of the mass of the particle. That is to say, that both the force on the particle and the inertia of the particle are proportional to the mass m, and thereby the frequency f, of the particle. So all particles under the influence of a gravitational field will irrespective of their mass be given the same acceleration, and thereby the same velocity.

As the gravitational force F_G is equal to $G(\varepsilon_0 \mu_0 h)^2 \cdot f_1 \cdot f_2 / r^2$, the gravitational force between two wave-particles is proportional to the product of their frequencies, and inversely proportional to the square of the distance between them.

Furthermore, as the vacuum permittivity ε_0 and the vacuum permeability μ_0 are included in the expression for the gravitational force F_G, gravity will be of an electromagnetic nature, which is why the gravitational force, just like the electromagnetic force, propagates with the speed of light in the ZPF.

2.4.5 Black Holes in the Euclidean Space

A black hole is a region of space in which the gravitational field is so strong that nothing - not even light - can escape from its pull. To calculate the escape velocity in a Euclidean space, we consider a heavy body of mass M, which is located at the origin of a coordinate system. Another body of mass m starts at the distance r from the origin with the velocity v. If the body were to escape to infinity, it must have a sufficient kinetic energy to be able to counterbalance the gravitational potential energy: [38]

$$\frac{mv^2}{2} = \frac{GMm}{r}. \tag{2.46}$$

For each value of v, there is a critical value of r, so a particle with velocity v is only able to escape to infinity, if

$$v^2 \geq \frac{2GM}{r}. \tag{2.47}$$

When the velocity is equal to the speed of light c_0, we get the radius of a black hole with the mass M from which nothing, not even light, can escape

$$r_{Schwarzschild} = \frac{2GM}{c_0^2}. \tag{2.48}$$

The value of the radius of a black hole is called the Schwarzschild radius. From the inequality (2.47) it can be seen, that if we set the velocity v equal to the velocity of light c_0, the gravitational field will not be able to hold on to matter, if the distance r becomes greater than $r_{Schwarzschild}$.

2.4.6 The Clocks go Slower in a Gravitational Field than Outside

As the wave-particles are changing under the influence of a gravitational field, it can be expected that the gravitational field also affects the mechanical clocks. If we use a stable oscillator as a clock and let its frequency represent a unit of time, the frequency will, as we have seen (eq. 2.40), go slower when the oscillator is affected by a gravitational field. This means that the clocks run more slowly in a gravitational field than in the free space.

The energy of a wave-particle can be expressed as $E = m_i c_0^2$, when it resides in free space without any external forces, and as E', when the wave-particle is in a gravitational field, where:

$$E' = m_i c_0^2 - \frac{GMm_i}{r} = E\left(1 - \frac{GM}{c_0^2 r}\right),$$

and $-GMm_i/r$ is the gravitational potential energy. Since a wave-particle contains the energy:

$$E = h \cdot f.$$

where f is the frequency of the emitted radiation, and h is the Planck constant, the relationship between the frequencies can be written as:

$$f' = \left(1 - \frac{GM}{c_0^2 r}\right) f \, .$$

Since the frequency is changed, and an oscillation represents a time unit, the clock will be slower. So we find the following connection between the clock time in a gravitational field t' and the clock time in the ZPF t:

$$t' = t\left(1 - \frac{GM}{c_0^2 r}\right) . \qquad (2.49)$$

So the clocks go slower, when they are in a gravitational field.

2.4.7 Energy and Mass are Deflected in a Gravitational Field

The existence of closed universes and black holes in a flat Euclidean space, depends on the fact that mass as well as energy are deflected in a gravitational field. This can be shown from the fact that the mass of a body m, is a function of the energy the body contains. If the growth of energy amounts to ΔE, the mass increases with the factor $\Delta m = \Delta E/c_0^2$. This means that if there is an increase of energy, there is a corresponding increase of the gravitational mass. [7]

An increase of the energy will then result in a corresponding increase in the potential energy in a gravitational field, $ghm = ghE/c_0^2$, where g is the gravitational acceleration, m is the mass, and h is the height. This applies to both energy and material bodies.

Using the masses of the photons we can calculate the deflection of electromagnetic radiation in a gravitational field, which stems from a massive body.

An electromagnetic radiation from a distant object will describe a characteristic hyperbolic path under the influence of the central force, when it passes the gravitational field from a celestial body. The deflections of the photons are shown in the figure, where, for the sake of understanding, the deflection is considerably exaggerated.

We designate the minimum distance between the radiation and the center of mass of the celestial body as r, the angle of the asymptote as φ, and the eccentricity of the hyperbola as e. Here, the connection between φ and e is equal to:

$$\cos\varphi = \frac{1}{e} \, , \quad where \; e = \frac{c_0}{a} \, .$$

The deflection angle of the radiation, α, is shown in the figure, and is:

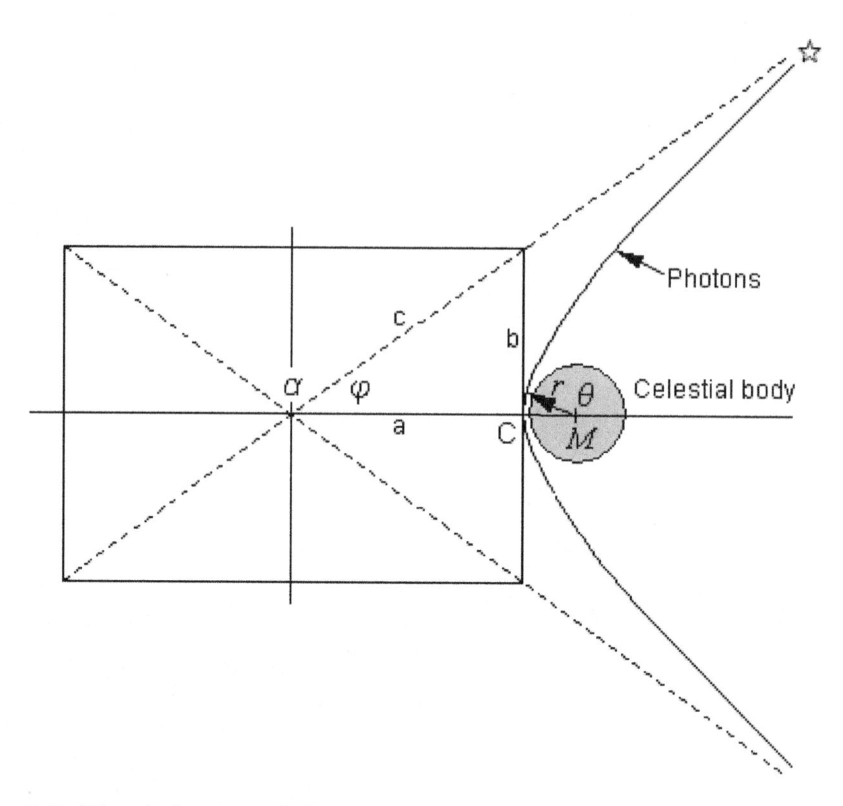

Figure 2.7: The deflection of electromagnetic radiation in a gravitational field.

$$\alpha = \pi - 2\varphi\,.$$

From the law of conservation of energy and angular momentum, we can find the total energy as a function of r:

$$E = \frac{1}{2}\mu\left(\frac{dr}{dt}\right)^2 + \frac{L^2}{2\mu r^2} - \frac{Gm_iM}{r}\,, \tag{2.50}$$

where $\mu = \frac{m_iM}{m_i+M} \approx m_i$ is the reduced mass of the particle, $\frac{1}{2}\mu(\frac{dr}{dt})^2$ is the radial kinetic energy, $\frac{L^2}{2\mu r^2}$ is the centrifugal potential, and $\frac{Gm_iM}{r}$ is the gravitational potential. [39]

A solution, to the radial energy equation, is:

$$r = \frac{p}{e\cos\theta + 1}\,,\ where\ p = \frac{L^2}{Gm_iM\mu}\,. \tag{2.51}$$

The total energy E is most easily found at the point C, where $\theta = 180^0$ and thus:

$$r = \frac{p}{1-e}, \quad and \quad \frac{dr}{dt} = 0\,.$$

If these terms are inserted in the energy equation, we get:

$$E = \frac{Gm_iM}{2p}(e^2 - 1)\,.$$

If we insert the expression for p and set $\mu \approx m_i$, we find:

$$E = \frac{G^2m_i^3M^2}{2L^2}(e^2 - 1)\,.$$

For an energy quantum, with mass m_i, the total energy E, and the angular momentum L with respect to the center of the celestial body, the eccentricity is thus equal to:

$$e = \left(1 + \frac{2EL^2}{G^2m_i^3M^2}\right)^{\frac{1}{2}}\,.$$

The constants of motion E and L can be easily obtained at the point C. If we at this point call the velocity v, and replace the reduced mass with m_i, we get:

$$E = \frac{L^2}{2m_ir^2} - \frac{Gm_iM}{r} = \frac{(m_irv)^2}{2m_ir^2} - \frac{Gm_iM}{r} = \frac{1}{2}m_iv^2 - \frac{Gm_iM}{r}\,,$$

where the angular momentum is equal to:

$$L = m_irv\,.$$

As, the velocity of the photons is equal to the speed of light c_0 - and thus, unless we are dealing with a black hole, much larger than the escape velocity $v_e = (2GM/r)^{\frac{1}{2}}$ (eq. 2.47) at the point C - we can ignore the change of the velocity of the photons and assume that the gravitational field only changes the direction of the velocity. If we set $v = c_0$, where c_0 is the velocity of the photons in the free space, the eccentricity can be written as:

$$e = \left(1 + \frac{(c_0^2 - 2GM/r)c_0^2r^2}{G^2M^2}\right)^{\frac{1}{2}}\,. \tag{2.52}$$

As anticipated, the masses of the photons cancel out. The value of the eccentricity determines whether the electromagnetic radiation can escape from the gravitational field of a celestial body or whether it will be captured. If $e \geq 1$, the mass of the celestial body will not be sufficiently large to hold on to the light, which thereby will describe a hyperbola

- or a parabola when $e = 1$.

It is seen that $e = 1$ when:

$$r = \frac{2GM}{c_0^2}. \tag{2.53}$$

This value of the radius is called the Schwarzschild radius, and if the radius is less than this value, then $e < 1$ and the electromagnetic radiation will describe an ellipse. The radiation will thus be unable to escape the gravitational field of the object. Here, the object can be a black hole or an entire universe. If it is a universe, it is referred to as a closed universe.

If we look at the cases where e is significantly larger than 1, and thus $2GM/r$ significantly smaller than c_0^2, the eccentricity can be simplified to:

$$e = \frac{c_0^2 r}{GM}.$$

The angle of deflection, of the radiation α can, by using the connections:

$$\alpha = \pi - 2\varphi,\ \cos\varphi = \frac{1}{e}\ \textit{and}\ e = \frac{c_0^2 r}{GM},$$

be written as:

$$\alpha = \pi - 2\arccos\left(\frac{GM}{e^2 r}\right). \tag{2.54}$$

Since $x = GM/(c_0^2 r) << 1$, $\arccos x$ can be developed in a Taylor series:

$$\arccos x = \frac{\pi}{2} - \arcsin x = \frac{\pi}{2} - \left(x + \frac{1}{2}\frac{x^3}{3} + \ldots\right).$$

As we only consider the first term within the brackets, the angle of deflection becomes:

$$\alpha = \frac{2GM}{c_0^2 r}. \tag{2.55}$$

Light passing along a body of mass M, will then be subjected to an angular deflection α, which also entails a reduction of the gravitational potential. In the expression for the angular deflection G is the gravitational constant, M is the mass of the celestial body, c_0 is the speed of light, and r is the distance between the light and the center of mass of the celestial body.

2.4.8 The Relativistic Gravitational Force

The Quantum Ether Theory yields an obvious explanation of the relativistic gravitational force. According to the QET, the gravitons are a part of the quantum field in which they propagate. Since the gravitons that establish the gravitational force between the masses, according to QET consist of virtual particles with the velocity c_0 relative to the ZPF, the massive objects, which have a velocity relative to the field, will also have a velocity relative to the gravitons that hold the masses together.

Let us make the calculation for the relativistic gravitational force. According to the gravitational law, the force between two masses m_1 and m_2, which are held together by the gravitational force and are at rest relative to the ZPF, is equal to:

$$F_G = G \cdot m_1 \cdot m_2 / r^2 \,,$$

where F_G is the gravitational force, r is the distance between m_1 and m_2, and $G = 6.674 \times 10^{-11}$ Nm^2/kg^2 is the universal gravitational constant.

Since we assume that the gravitational force propagates in the ZPF with the speed of light, c_0, the distance r can be written as $c_0 t$, where t is the time it takes the forces to move the distance r, i.e. $r = c_0 t$.

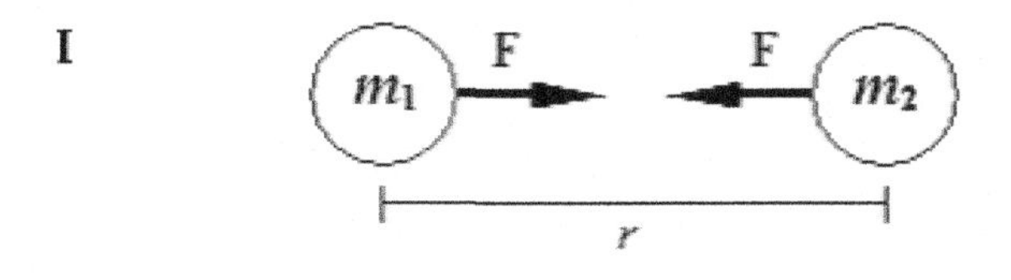

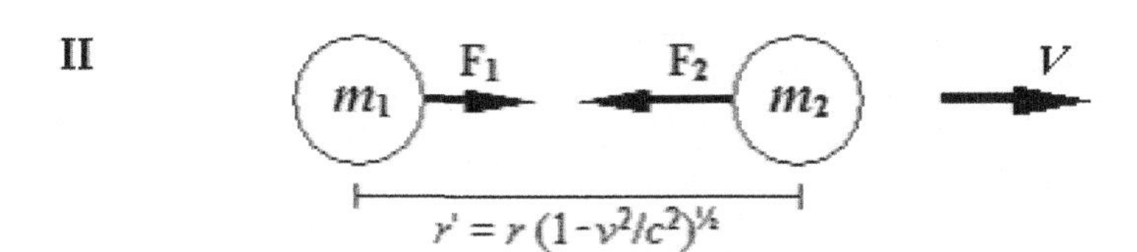

The two masses are then set in motion in the direction from m_1 to m_2, with the velocity v relative to the ZPF. Since both the forces, as well as the distance between the masses, may have changed as a result of the movement relative to the field, we write:

$$F'_G = G \cdot m'_1 \cdot m'_2 / r'^2 \,.$$

In the calculation, the masses m_1 and m_2 are assumed to be constant whether or not they are moving relative to the ZPF. Since the forces propagate at the speed of light c_0

in the ZPF, the distance the force has to travel from m_1 to m_2 becomes to the first order in vt: $r_1 = c_0 t + vt$ (since the mass m_2 moves away from m_1 with the velocity v relative to the ZPF). The distance the force has to travel from m_2 to m_1 becomes to the first order in vt: $r_2 = c_0 t - vt$ (since the mass m_1 approaches m_2 with the velocity v relative to the ZPF).

We thus find:

$$F'_G = Gm_1m_2/r'^2 = Gm_1m_2/[r_1r_2] = Gm_1m_2/[(c_0t + vt)(c_0t - vt)] = Gm_1m_2/[c_0t(1 + v/c_0)c_0t(1 - v/c_0)] \,,$$

where from we find the relativistic gravitational force:

$$F'_G = Gm_1m_2/[r^2(1 - v^2/c_0^2)] \,. \tag{2.56}$$

This is the mutual gravitational force that two masses experience when they move with the common velocity v relative to the ZPF along the line that connects the two masses. So the gravitational force will increase with the factor $\gamma = (1 - v^2/c_0^2)^{-1}$.

Since the gravitational force F_G can be expressed as $F_G = G(\varepsilon_0\mu_0 h)^2 \cdot f_1 \cdot f_2/r^2$, where G is the gravitational constant, f_1 is the frequency of the wave packet related to the first particle, and f_2 is the frequency of the second wave packet. The relativistic gravitational force F'_G can also be expressed as:

$$F'_G = G(\varepsilon_0\mu_0 h)^2 \cdot f_1 \cdot f_2/[r^2(1 - v^2/c_0^2)] \,. \tag{2.57}$$

The gravitational force we have found here is true when the particles have the mutual spacing r and move along their connecting line. As the particles move with their individual phase velocities, there is nothing to prevent the particles from moving faster than the velocity of light (eq. 2.16). When this occurs, the relativistic gravitational force will change sign, so it becomes repulsive.

2.5 The Redshift

A redshift occurs when electromagnetic waves undergo a shift towards the less energetic end of the electromagnetic spectrum either due to the Doppler effect, a gravity displacement, or a plasma redshift. The redshift leads to an increase in the wavelength of the electromagnetic radiation compared to the wavelength originating from the source. This increase in wavelength corresponds to a decrease of the frequency, since:

$$\lambda \cdot f = c_0 \,,$$

where λ denotes the wavelength, f the frequency, and c_0 the speed of light. A shortening of the wavelength is called a blueshift.

The redshift or blueshift can be characterized by the relative difference between the observed and emitted wavelength (or frequency). In astronomy, it is customary to describe this change, by using the dimensionless size z, where the redshift is defined as:

$$z = \frac{\lambda_r - \lambda_s}{\lambda_s} \; or \; 1 + z = \frac{\lambda_r}{\lambda_s} \, .$$

Here, λ_r is the received wavelength, and λ_s is the emitted wavelength. These equations can also be expressed by use of the frequencies:

$$z = \frac{f_s - f_r}{f_r} \; or \; 1 + z = \frac{f_s}{f_r} \, . \tag{2.58}$$

The sign of z determines whether there is a redshift or a blueshift. If the light source moves away from the observer, escapes from a gravitational field, or penetrates a hot, sparse electron plasma, [42] there is a redshift, and $z > 0$. When the source moves towards the observer, it will result in a blueshift, and $z < 0$.

2.5.1 The Plasma Redshift

The intergalactic space consists of a hard vacuum that contains a plasma of electrons, hydrogen and helium ions, as well as electromagnetic radiation, magnetic fields, neutrinos, dust, gas, and cosmic rays. The density of the hydrogen ions is roughly one atom per cubic meter. [43], [44] However, despite the rarity of the particles the temperature never gets lower than the temperature of the CMB radiation at 2.7 K, [45] while the temperature of the plasma, which contains 40–50% of the baryonic matter of the Universe, is as high as 10^5 to 10^7 K. [46] This plasma is also called the warm–hot intergalactic medium. Otherwise, plasma is mostly associated with the interior of the Sun, [47] the solar corona, [48] the stars, the galaxies and the quasars. [49]

Plasma or ionized gas has unique properties and displays a behavior different from that of the other physical states. It is a gaseous mixture of negatively charged electrons and charged positive ions, which is created by heating or by a strong electromagnetic field. This distinct separation of ions and electrons produces an electric field, which in turn produces electric currents and magnetic fields. When light traverses the sparse hot plasma, the light becomes red-shifted, rather than being scattered. This can be explained by the fact that the electromagnetic field from the electrons gradually removes tiny portions of the energy from the electromagnetic photons. It is the same phenomenon that takes place in the intergalactic medium, a planetary nebula and a corona. That is to say, the warming of the Sun's corona is evidence that the radiation from the interior of the Sun loses energy to the corona's plasma when the radiation passes through it, causing the radiation to redshift.

The plasma redshift is a multiple scattering process where each penetrating photon interacts with a great many electrons in the plasma. At the plasma redshift, the penetrating photons with a wavelength of 500 nm transfers an energy of about $3.3 \times 10^{-25} hf$ to the plasma per electron, [42] so the heating during a plasma redshift is very large and important for explaining the heating of the solar corona, the corona of galaxies, and the intergalactic plasma. When the photons are in the far infrared, the cross section becomes very large and approaches infinity when the infrared photons' energy approaches zero, so when the photons are in the far infrared, whereby they are very soft, the interaction in a hot, sparse plasma involves many electrons, even in the sparsest plasma of intergalactic space. [50], [51]

Within reasonable boundaries all the electrons have different energy levels, so the quantum numbers of angular moments in the interactions between the electrons are large. The cross section for the plasma redshift depends on the photon width and the damping in the plasma. The damping in the electron plasma varies with the plasma temperature and how the damping and the density of the electrons affect both the coherence effects and the cross section for the plasma redshift of the photons. The plasma redshift varies with the wavelength, the electron temperature and the density. Only when the wavelength is less than a certain cut-off wavelength, which depends on the electron temperature and the density, is the plasma redshift significant. However, the plasma redshift and the cut-off wavelength for the redshift is also affected by the magnetic field.

As the electrons keep each other at a distance in the hot, sparse plasma, the cross section for each individual interaction is very large, so any detectable effect involves a large number of interactions with many electrons. The energy that is removed in this way is very tiny per electron; however, when the energy is removed little by little, the light becomes red-shifted, as the light traverses the vast distances of sparse plasma. This plasma redshift is only efficient when the photons penetrate a hot, sparse electron plasma such as the solar corona, the corona of stars and galaxies, the interstellar and intergalactic plasma, and the plasma of quasars. However, when photons penetrate a cold and dense electron plasma, they lose energy through ionization, excitation, and Compton scattering on the individual electrons, and through Raman scattering on the plasma frequency. [42]

A Poynting vector S represents the directional energy transfer per unit area per unit time of an electromagnetic field, which often is measured in watts per square meter ($\mathrm{W{\cdot}m^{-2}}$). For a field of photons moving along the x-axis, the Poynting vector S can be normalized to the energy flux of one photon, $\hbar\omega_0 = hf_0$, per second per cm^2 in vacuum, where $\hbar$ is the reduced Planck constant $h/(2\pi)$, and $\omega_0 = 2\pi f_0$ is the cyclic frequency of the incident photon with the frequency: f_0.

Even in a vacuum the photon is never infinitely sharp, but consists of a distribution of frequency components as indicated by:

$$S = \hbar\omega_0 = \hbar\omega_0 \frac{\gamma}{2\pi} \int_{-\infty}^{\infty} \frac{d\omega}{(\omega - \omega_0)^2 + \gamma^2/4}, \tag{2.59}$$

where γ is the photon width. [50] When the photon penetrates a plasma the photon's virtual field will be modified by the dielectric constant $\epsilon = (n - i\kappa)^2$, where n is the refraction index and $i\kappa$ the absorption coefficient. When the binding-energy frequency of the electron $\omega_q = 0$ and the collision damping $\alpha = 0$ (because the collision damping, α, is included in $\beta\omega^2$), we find that the dielectric constant is:

$$\epsilon = (1 - \frac{\omega_p^2}{\omega^2 + \beta^2\omega^4}) - i\frac{\beta\omega\omega_p^2}{\omega^2 + \beta^2\omega^4}, \tag{2.60}$$

where $\omega_p = \sqrt{4\pi e^2 N_e / m_e}$ is the plasma frequency, e the electron charge, N_e is the number of plasma electrons per cm^3, m_e is the mass of an electron, and $\beta\omega^2$ is the radiation damping in the hot, sparse plasma. Since the damping factor, β, changes with the temperature, the density of the plasma, and the wavelength of the incident light, the collective effects are very large, so the plasma-redshift becomes significant. [42]

When the photons lose energy during the plasma redshift, the photons transfer energy to the plasma in extremely small quanta. In the quiescent solar corona, this heating starts in the transition zone to the solar corona and is a major fraction of the coronal heating. Generally, the part of the photon energy that is transferred to the plasma causes a significant heating of the plasma, so the plasma redshift contributes to the heating of the interstellar plasma, the solar corona, the corona of galaxies, and the intergalactic plasma. In this way, the plasma redshift explains the solar redshifts, and the redshifts of galactic coronas, and leads to a hot intergalactic plasma, which can explain the cosmological redshift and the microwave background, and the X-ray background. [42]

Moreover, it has been shown that the plasma redshift can explain the observed magnitude-redshift relation for Ia-supernovae, and that the Ia-supernovae observations are in better agreement with the redshift predicted by the plasma redshift, than that predicted by the big-bang cosmology. [42]

If the redshift is interpreted as a velocity with direction away from the observer, the measured red-shift corresponds to an expansion of the Universe. However, no form of energy exists capable of producing such an expansion. But as the measured redshift can be explained as a combination of the plasma redshift, the Doppler shift, and the gravitational redshift, this problem does not arise. Since the universe does not expand, it will be Euclidean and quasi-static on the large scale, without Einstein's Lambda expression. So the distribution of matter and energy in the flat space is determined by the fundamental forces and the mass distribution existing at any given time.

2.5.2 The Doppler Effect

The Doppler effect denotes the phenomenon that the frequency or wavelength of a wave changes in relation to an observer, depending on the velocity of the source relative to the observer. Since waves according to the QET propagate in the zero-point field, such as for instance electromagnetic waves, the velocity of the observer and the source can be measured relative to the media in which the waves propagate. The total Doppler effect is then a result of the motion of the source and the observer relative to the field.

The relationship between the observed frequency f_r (receiver) and the emitted frequency f_s (source) is given by:

$$f_r = \frac{v + v_r}{v + v_s} \cdot f_s \,, \tag{2.61}$$

where v is the velocity of the wave in the medium, v_s is the velocity of the source relative to the medium, and v_r is the velocity of the recipient relative to the medium. [52]

The sign of the velocities v_s and v_r must be selected, so that the observed frequency increases when the source is moving towards the observer or the observer moves towards the source. This means that the sign of the velocity v_s must be positive when the source moves away from the observer, and the velocity v_r must be negative when the observer moves away from the source. The formula assumes that the source is moving directly towards or away from the observer.

In the cases where the velocity of the wave is significantly larger than the relative velocity of the source or the observer, as is usually the case with electromagnetic waves with the velocity c_0, the ratio between the observed frequency f_r and the emitted frequency f_s can be written as:

$$f_r = \frac{c_0 + v_r}{c_0 + v_s} \cdot f_s \approx (1 - \frac{v_{s,r}}{c_0}) \cdot f_s \,, \tag{2.62}$$

where $v_{s,r}$ is the velocity of the source relative to the receiver, and c_0 is the speed of light in the ZPF. The size $v_{s,r}$ is negative when the source moves towards the receiver and positive when it moves away from the receiver.

The Doppler effect results from the relative motion of a light-emitting object and an observer. If the source of light is moving away, then the wavelength of the light is stretched out; that is to say, the light is shifted towards the red. If the source is moving towards us, the light is shifted towards the blue. These effects, individually called the redshift and the blueshift, are together known as Doppler shifts.

If the source moves away from the observer with the velocity v_s, the Doppler effect can be found by (eq. 2.58):

$$1 + z = \frac{f_s}{f_r}, \text{ where (eq. 2.62) } \frac{f_s}{f_r} = \frac{c_0 + v_s}{c_0 + v_r} \approx \frac{c_0 + v_s}{c_0} = 1 + \frac{v_s}{c_0},$$

from which the redshift z, due to the Doppler shift, is:

$$z = \frac{v_s}{c_0}, \tag{2.63}$$

where c_0 is the speed of light. The formula holds as long as the velocity v_s is significantly less than the speed of light.

2.5.3 The Gravitational Redshift

Light and other forms of electromagnetic radiation that originate from a source located in a strong gravitational field will have a longer wavelength than radiation emitted from a source located in a region of a smaller gravitational field. This is because photons that leave a gravitational field lose an internal amount of energy equal to the potential energy they must overcome to escape from the strong field. Since the lower energy at the long-wave end of the visible electromagnetic spectrum is red, the extension of the wavelength is called a redshift.

A blueshift, on the other hand, is a shortening of the wavelength of the emitted radiation or an increase in the frequency of the radiation. The name derives from the fact that the shorter end of the visible spectrum is blue or violet.

Redshift and blueshift can be derived by considering the energy outside and inside a gravitational field. Let us consider the energy of a photon with the mass m that stays in an area which is not affected by a gravitational field, and let us denote the energy of the received photon with E_r. Since the photon has not been affected by any external forces, an observer will measure the energy:

$$E_r = mc_0^2. \tag{2.34}$$

If the same photon has escaped from a position r inside the gravitational field of a heavy body with the mass M, it must have had an additional energy equal to the energy it needed to escape from the gravitational field, which must be numerically equal to the gravitational potential energy at the position r, but with an opposite sign. So, it would have needed the additional energy:

$$\frac{GmM}{r},$$

to escape from the gravitational field of the heavy body, where G is the gravitational constant and r is the distance between the photon and the center of mass of the heavy body. The total energy of the photon E_s must then have been equal to:

$$E_s = mc_0^2 + \frac{GmM}{r}, \tag{2.64}$$

when it resided in the gravitational field of the heavy body. Because the energy of the photon is a function of its distance r from the center of mass of the heavy body, the energy E_s can be written as:

$$E_s = mc_0^2\left(1 + \frac{GM}{c_0^2 r}\right) = E_r\left(1 + \frac{GM}{c_0^2 r}\right). \tag{2.65}$$

According to the Quantum Theory, the energy of a quantum of radiation is equal to Planck's constant h times the frequency f,

$$E = h \cdot f. \tag{2.35}$$

From this connection, we can find the frequency f_s of the photon in the gravitational field expressed by the received frequency f_r outside the heavy body:

$$f_s = f_r\left(1 + \frac{GM}{c_0^2 r}\right). \tag{2.66}$$

As previously mentioned (equation 2.58) we have that the redshift equals:

$$z = \frac{f_s}{f_r} - 1, \tag{2.58}$$

from which we find the gravitational redshift:

$$z \approx \frac{GM}{c_0^2 r}, \tag{2.67}$$

where r is the distance from the center of mass, from which the photon escapes the celestial object with the mass M to a point "outside" the gravitational field of the object. To gain an overview of the gravitational redshift, we calculate the gravitational redshift generated by the Sun and by a neutron star.

As the gravitational constant is equal to $G = 6.67 \times 10^{-11}$ m^3 kg$^-1$ s$^-2$, the speed of light $c_0 = 3.00 \times 10^8$ m/s, the mass of the Sun $M_\odot = 1.99 \times 10^{30}$ kg, and the radius of the Sun $R_\odot = 6.96 \times 10^8$ m, we find, that the gravitational redshift generated by the Sun is:

$$z = \frac{GM_\odot}{c_0^2 R_\odot} = \frac{6.67 \times 10^{-11} \cdot 1.99 \times 10^{30}}{(3.00 \times 10^8)^2 \cdot 6.96 \times 10^8} = 2.1 \times 10^{-6}\,,$$

and as the mass of a regular neutron star is $M = 4.0 \times 10^{30}$ kg, and the radius is $r = 1.0 \times 10^4$ m, the gravitational redshift generated by a neutron star is:

$$z = \frac{GM}{c_0^2 r} = \frac{6.67 \times 10^{-11} \cdot 4.0 \times 10^{30}}{(3.00 \times 10^8)^2 \cdot 1.0 \times 10^4} = 0.3\,.$$

Because our solar system is located within the Milky Way, the blueshift from our position in the Milky Way will to some extent offset the redshift from the sources situated outside our galaxy.

2.5.4 The Cosmological Redshift

The light from distant stars and even more distant galaxies is not featureless, but has different spectral features, characteristic of the atoms in the gases around the stars. When these spectra are examined, they are mainly shifted toward the red end of the spectrum because of a combination of the plasma redshift, the Doppler effect, and the gravitational redshift.

The redshift, z, is defined as the change in the wavelength of the light divided by the rest wavelength of the light:

$$z = \frac{\lambda_{obs} - \lambda_{rest}}{\lambda_{rest}} = \frac{f_{rest} - f_{obs}}{f_{obs}}\,.$$

Here λ_{obs} and f_{obs} are the observed wavelength and frequency, and λ_{rest} and f_{rest} are rest values in relation to the ZPF. Note that positive values of z correspond to redshifts, while negative values of z correspond to blueshifts.

A. *The Plasma Redshift*
The plasma redshift accounts for the loss of energy of the photons, when they move through a hot, sparse electron plasma from the source to the observer.

B. *The Doppler Effect*
The Doppler effect denotes the phenomenon that the frequency or wavelength of a wave change in relation to an observer, depending on the velocity of the source relative to the observer.

C. *The Gravitational Redshift*
The gravitational redshift is a shift in the frequency of a photon to a lower energy when it climbs out of a gravitational field. This gravitational redshift can be compensated by a shift in the frequency of the photon to a higher energy when it enters a gravitational field; and because our solar system is placed inside the Milky Way, the blueshift, which originates from our position in the Milky Way, will to some extent counterbalance the redshift from sources placed outside our galaxy.

By comparing the three types of red-/blueshifts, it can be seen that the Doppler effect and the gravitational redshift are indifferent to the distance from the observer and dominate at the cosmological vicinity, while the plasma redshift dominates at larger distances.

2.5.5 Hubble's Law

Hubble's law is the name of the astronomical observation that all objects observed in deep space are found to have a redshift proportional to their distance from the Earth. [53]

The assumption that the redshift could be interpreted as the velocity of the source led to the idea that the speeds of the receding galaxies are proportional to their distance from the Earth, so Hubble's law could be given the simple mathematical expression:

$$v_s = H_0 D_H,$$

where v_s is the speed of the source in agreement with the redshift, H_0 is Hubble's constant, and D_H is the distance from the source to the observer. But where should the energy come from to generate such an expansion?

2.5.6 Hubble's Constant

The Hubble constant H_0, which has been interpreted as the expansion rate of the universe, can be expressed as a product of the redshift z (eq. 2.63) and the inverse Hubble time t_H, where the Hubble time is assumed to be the actual age of the Universe:

$$H_0 = \frac{v_s}{D_H} = \frac{v_s}{c_0 \cdot t_H} = \frac{v_s}{c_0} \cdot \frac{1}{t_H} = z \cdot \frac{1}{t_H}.$$

But since the redshift according to the Quantum EtherTheory is due to a combination of the plasma redshift, the Doppler shift, and the gravitational redshift, the Hubble constant H_0 times the Hubble time just describe the combination of these redshifts.

The plasma redshift is a result of the total path of the photons through the hot sparse plasma, during which the photons transfer some of their energy to the plasma. The plasma redshift occurs when photons penetrate a hot, sparse electron plasma, where the photons lose energy to the plasma, in an interaction with the electrons.

The energy loss of the photons consists of very small quanta, which are absorbed by the plasma and cause a significant heating of the solar corona, the interstellar plasma, the galactic corona, and the intergalactic plasma. This explains the CMB, the X-ray background, the solar redshifts, the redshifts of the galactic corona, and the cosmological redshift. In this way, the redshift supersedes the cosmological redshift that describes the lengthening of the wavelength of the emitted radiation due to a hypothetical expansion of the Universe. [42]

The Doppler shift and the gravitational redshift explain the deviations from the cosmological redshift. The Doppler shift explains the deviations from the cosmological redshift that arise from the peculiar motions by galaxies and clusters, while the gravitational redshift (eq. 2.67) is due to a lengthening or shortening of the wavelength depending on whether the photons climb out of or enters a gravitational field. [54]

The Hubble constant H_0 is not an expansion parameter, but is instead a measure of the average electron density along the line of sight towards an object. So the Universe does not expand, and measurements of the absolute magnitudes and redshifts of Ia supernovae show that the plasma redshift can explain the observed redshifts with a greater accuracy than Hubble's constant. [55]

Consequently, the universe does not expand, but remains in a quasi-static state, which is determined by the variations in the gravitational field and the distribution of matter and energy - such as black holes, the warm–hot intergalactic medium, stars, supernovae, active galactic nuclei (AGNs) and radiation.

Chapter 3

The Euclidean Cosmos Theory

3.1 Deduction of the Structure and Composition of the Cosmos

When we consider the Universe, we tend to look at the Universe from our own limited perspective. However, the age of our Universe at 13.8 billion years [56] is nothing compared to the infinite timescale, and the visible extent of the Universe, which is equal to the current horizon distance $d_{hor}(t_0) \approx 14.3$ Gpc or 4.4×10^{26} m, [57] is nothing compared to infinity.

When we present a theory of the distribution of matter and energy in the Cosmos, it is essential that we start with the observations that are made in our own Universe. Moreover, we must begin with the physical laws that have proved viable when tested in relation to the world that surrounds us.

In the following deduction of the energy distribution in space, we assume that the Cosmos has existed infinitely, that the energy is constant, that the space is Euclidean and thus completely flat, and that the matter and energy are quantized – and therefore cannot end as a singularity. The gravitational forces will then produce a mass distribution in the infinite flat space, where the matter and energy will accumulate into larger and denser structures, until a state of equilibrium in the Euclidean space arises.

As time goes by, the larger and denser structures will accumulate in black holes and closed universes, and since quantum theory does not allow singularities, even the closed universes will end up as gigantic black holes when the energy becomes exhausted. But since we exist, there must be a way out - there must be a way in which a black hole can be converted into energy. That is to say, a black hole must be able to create an explosion where $E = mc_0^2$.

We now present a more formal deduction of the outlined course of events, from the following assumptions:

1. ***The law of conservation of energy.*** [11]
2. ***The space is Euclidean.***
3. ***No EM interactions move faster than the speed of light in vacuum.***
4. ***Matter and energy are deflected in a gravitational field.***
5. ***The Cosmos has existed for an infinitely long time.***
6. ***We exist.***

3.1.1 Assumptions for the Determination of the Energy Distribution

Ad 1) The law of conservation of energy.
The first assumption is simply the first law of thermodynamics, or the law of conservation of energy, that states that: Energy can neither be created nor destroyed, but only be changed from one form to another. [11]

In this connection it is essential to mention that since all matter and energy are quantized, with the minimum length called the Planck length, the existence of a singularity is at variance with the quantum field theory. This is because, when the dimensions of the singularity approach zero, the dimensions become less than the Planck length, which means that the singularity cannot contain as much as a single quantum of energy.

Ad 2) The space is Euclidean.
Since it has been established that time is absolute and universal (ch. 2.3.4), the time axis is as linear as the three space axes (x, y, z), so the combined space-time can best be interpreted as a Euclidean space with three space axes and one time axis.

Ad 3) No EM interactions move faster than the speed of light in vacuum.
Since it has been found that the gravitational force is equal to $F_G = G(\varepsilon_0\mu_0 h)^2 f_1 f_2/r^2$ (eq. 2.45), it will be of a electromagnetic nature, which is why gravity, just as the electromagnetic force, propagates with the speed of light in vacuum. This has also been established by the quantum field theory. [12]

The reason why the speed of the electromagnetic radiation is constant and independent of the velocity of the object emitting the radiation is that electromagnetic radiation propagates in the ZPF with the constant speed of light:

$$c = \frac{1}{\sqrt{\varepsilon_0\mu_0}},$$

where ε_0 is the electric constant and μ_0 is the magnetic constant. However, that the speed of light is constant does not mean that light cannot be deflected in a gravitational field.

Ad 4) Matter and energy are deflected in a gravitational field.
Since mass and energy are equivalent entities, where $E = mc_0^2$, (eq. 2.34), and since we from Newton's law of universal gravitation know that the mass m is deflected in a gravitational field, it must also be true for the energy E. If we look at a closed universe, it will thus not be the space that bends, but the trajectories of matter and energy that are deflected in the gravitational field. This is true for all types of electromagnetic radiation including light.

Ad 5) The Cosmos has existed for an infinitely long time.
The fifth condition is a result of the Euclidean geometry of space and the law of conservation of energy.

Ad 6) We exist.
The assumption that we exist ensures that at least one universe exists.

3.1.2 Deduction of the Energy Distribution in the Cosmos

The theory is a logical deduction of the composition of the Cosmos based on the given assumptions:

1. Assumption: The law of conservation of energy.
The energy is constant. [11] $\Rightarrow$ **The total amount of matter and energy is finite.**

Comment: Since energy can neither be created nor destroyed according to the first assumption, the total amount of energy in the Cosmos is constant. If the total amount of energy is constant, the amount of matter and energy will have a maximum, so the quantity of matter and energy is finite.

2. Assumption: The space is Euclidean.
The total amount of matter and energy is finite. $\wedge$ **The space is Euclidean.**
$\Rightarrow$ **The finite amount of matter and energy can be enclosed by a hypothetical spherical shell in the Euclidean space.**

Comment: If the space is Euclidean, only one connected space exists in which the finite amount of matter and energy must necessarily be. Since the quantity of matter and energy is finite, it will have a finite extent. This means that the amount of matter and energy can be enclosed by an outer boundary, so that we can place a hypothetical spherical shell around the finite amount of matter and energy.

3. Assumption: No EM interactions move faster than the speed of light in vacuum.
The finite amount of matter and energy in the Euclidean space can be enclosed by a hypothetical spherical shell. $\wedge$ **No EM interactions move faster**

than the speed of light in vacuum.
⇒ **The hypothetical spherical shell needed to enclose the finite amount of matter and energy grows at most at the speed of light in vacuum in the Euclidean space.**

Comment: During extreme conditions, such as in the explosion of black holes, the pressure and temperature can become so high that a plasma of free quarks and gluons can be produced which can reach speeds up to $\sqrt{3}c_0^2$ (ch. 3.4.7). However, when the pressure and temperature falls the quark-gluon plasma will immediately combine to create a jet of photons, nucleons and leptons.

4. Assumption: Matter and energy are deflected in a gravitational field.
Since matter and energy are deflected in a gravitational field, there are two possibilities, either: a) the finite amount of matter and energy cannot escape the gravitational field, that is, "**the universe is closed**", or b) the finite amount of matter and energy can escape the gravitational field, by which "**the universe is open or flat**".

4a. **The hypothetical spherical shell that is needed to enclose the finite amount of matter and energy grows at most at the speed of light in vacuum in the Euclidean space.** ∧ **The universe is closed.**
⇒ **The finite amount of matter and energy can be regarded as one closed universe, which is situated in the Euclidean space.**

Comment: If the density of matter and energy is sufficient to hold on to matter and energy, a constant spherical shell will be able to enclose the total amount of matter and energy. This means that the Euclidean space contains one closed universe.

4b. **The hypothetical spherical shell that is needed to enclose the finite amount of matter and energy grows at most at the speed of light in vacuum in the Euclidean space.** ∧ **The universe is open or flat.**
⇒ **The spherical shell that is required to enclose the finite amount of matter and energy grows at a velocity that is greater than zero and less than or equal to the speed of light in vacuum in the Euclidean space.**

Comment: If the density of matter and energy is not sufficient to hold on to matter and energy, the radius of the spherical shell will grow with a velocity that is greater than zero and less than or equal to the speed of light, so as to constantly encircle the total, constant amount of matter and energy in the Euclidean space.

5. Assumption: The Cosmos has existed for an infinitely long time.
The spherical shell that is required to enclose the finite amount of matter and energy will grow with a velocity that is greater than zero and less than or equal to the speed of light in vacuum in the Euclidean space. ∧ **The Cosmos**

has existed for an infinitely long time.
⇒ The density of matter and energy in the Euclidean space will approach zero with the exception of a finite number of accumulation points around which matter and energy are collected.

Comment: It must be true that a point can only be an accumulation point when matter and energy around the point of accumulation can create a gravitational force that is strong enough to hold on to matter and energy. So, when the density of matter and energy in the Euclidean space gather around an accumulation point, it will be either a barren object, a black hole, or a closed universe, since any form of energy otherwise would have radiated away long ago. So the conclusion can be reformulated as:

The density of matter and energy in the Euclidean space will approach zero with the exception of a finite number of barren objects, black holes, and closed universes.

6. Assumption: We exist.
The density of matter and energy in the Euclidean space will approach zero with the exception of a finite number of barren objects, black holes, and closed universes. $\wedge$ We exist.
⇒ In the Euclidean space a finite number of barren objects, black holes, and closed universes exist, and at least one closed universe.

Since the number of objects is finite, they can be enclosed by a hypothetical spherical shell. Within the hypothetical spherical shell, an individual barren object, black hole, or universe may at some time, either be in (or come into) possession of the escape velocity relative to the other objects, whereby the barren object, black hole, or universe will be thrown away from the other objects and become independent. For the barren objects, black holes, and universes, whose velocities never reach the escape velocity, it must be true that they because of the gravitational forces between them, will gather in one or more bounded areas. As the Cosmos has existed for an infinite length of time, the bounded areas must find themselves in relatively stable, dynamic equilibrium.

3.1.3 The Conclusion of the Energy Distribution in the Cosmos

We can finally conclude that the Cosmos consists of an infinite Euclidean space in which there is a finite number of closed universes, black holes and barren objects - and at least one closed universe. If there are more closed universes, black holes, and barren objects, they will either move away from each other at speeds that for each of them are higher than the escape velocity from the overall system, or be in a kind of stable, dynamic equilibrium, which means that there is a potential for barren objects, black holes and universes

to collide.

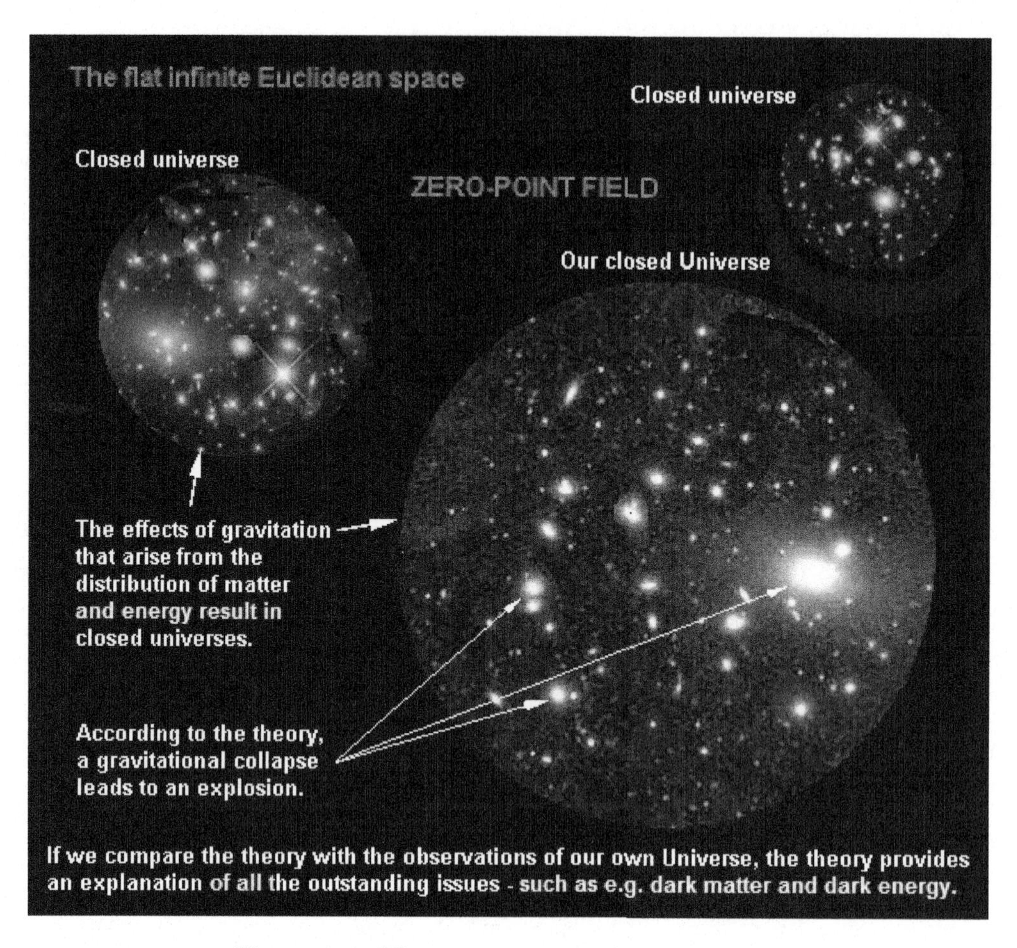

Figure 3.1: The composition of the Cosmos.

Comment: In honor of those who can imagine an infinite and (at the same time) constant amount of matter and energy, we will let the amount of matter and energy approach infinity. According to the theory, this can end in two scenarios. If the density of matter and energy is relatively small, the Cosmos consists of an infinite Euclidean space in which there is an infinite number of closed universes, black holes and barren objects - all of which are in a kind of stable, dynamic equilibrium, meaning that there is a potential for barren objects, black holes and universes to collide. However, if the Cosmos consists of a single coherent infinite Euclidean Universe, the density cannot be larger than the plasma redshift is capable of absorbing so much energy that the Universe does not suffer the heat death, as described by Olber's paradox.

3.2 The Energy Distribution in our own Universe

We will now look at what takes place in the closed universes. We will start from our own closed Euclidean Universe, as it is the only universe we know, but the conclusions we draw will apply to all the universes. From the observations of our Universe, we can see that matter and energy are assembled in galaxies, which in turn are assembled in super clusters, large quasar groups, galaxy filaments, galaxy walls and galaxy sheets.

Figure 3.2: Our closed Universe.

If we assume that universes generally have a content of hydrogen and helium similar to our own universe, which contains about 75% hydrogen [58] and 8% helium [59] of the total baryonic mass, the galaxies will go through a cycle of phases where the hydrogen and helium gather into nebulae, which again become stars, and then giants, white dwarfs, supernovae, neutron stars, and black holes. However, if the black holes were the final stage, and thereby not able to spread their content of matter and energy as AGNs, all the galaxies would ultimately end up as black holes. If that was the end result, the black holes would gravitate to the densest regions of the universe, which would act as accumulation points for the creation of still larger and heavier black holes, which would ultimately gravitate to the center of mass of the closed universe, to create one large black hole.

This is because even if there were only a few black holes left, which circulated around each other, they would ultimately merge into one single black hole. This is due to the loss of energy in the form of gravitational waves, where the gravitational waves in a gravitational field are equivalent to electromagnetic waves in an electromagnetic field. [60]

If black holes were the final stage of the universes, we would not be here. However, since we exist, there must be a process which is capable of generating an explosion of a black hole, and thereby spread its contents as radiation and matter.

3.2.1 A Gravitational Collapse must lead to an Explosion

When we consider a closed universe, matter and energy must, because of the gravitational force, amass in still larger black holes, which eventually will end up as one giant black hole. However, since we exist there must occur intermittent emissions of energy from the black holes. Such emissions will be called "regenerative processes", since the Cosmos without them would have been reduced to black holes and barren objects, infinitely long ago.

Larry Smarr calculated the first numerical solution to a direct collision between two black holes of equal masses in 1979, [61] - while Matzner and associates in 1995 determined the details for the merger. [62] The calculation of the direct collision of two black holes of equal masses, both of which start at rest, shows that when black holes fall against each other, they will merge to form one big black hole. At the beginning, the new black hole fluctuates, but as the oscillations die away, the hole settles down as a single spherical symmetric black hole. [63]

When more and more material is added to a black hole, the pressure rises at the center, and since a black hole from the outset has a mass that is so big that not even light can escape, all the matter and energy that is fed into the black hole will stay there. When a black hole gradually gets bigger, the material of the black hole goes through different phases depending on its distance from the center of the black hole.

That the masses of black holes can be added together can also be seen from the observations of black holes that consist of millions of solar masses. [64] There have even been observed super massive black holes (SMBHs) with masses around 40×10^9 solar masses. [65] and the greater a black hole becomes, the more matter and energy it will be able to attract, so the center of the black hole becomes denser as time passes by, until it becomes an AGN.

3.2.2 The Physical Conditions for the Creation of an Explosion

At ordinary temperatures quarks are bound together by gluons, whereby they make up the hadrons, which are divided into two families, the baryons and the mesons. The baryons, which each consist of three quarks, make up ordinary matter, such as protons and neutrons that together with the electrons form the atoms. The electrons themselves belong to a group of subatomic particles called leptons. [66]

In particle physics, which is the branch of physics that deals with fundamental particles and their properties, a fermion is any particle characterized by Fermi–Dirac statistics, while a boson is a particle that obeys Bose–Einstein statistics. [67] Fermions include leptons, such as the electron, and quarks, as well as any composite particles made up of an odd number of quarks, including baryons such as neutrons and protons. Fermions differ in general from bosons, in the manner that particles with half-integer spin are fermions, while particles with integer spin are bosons. Fermions obey the Pauli exclusion principle, which is the quantum mechanical principle that states that two or more identical fermions cannot simultaneously occupy the same quantum state within a quantum mechanical system. [68] In particular, it means that free identical fermions, that are limited to a finite volume, can only take on a discrete set of energies called quantum states. At the lowest total energy (when the thermal energy of the particles is negligible), all the lowest energy quantum states are filled. Such a state is referred to as full degeneracy, and its pressure, called the degeneracy pressure or Fermi pressure, remains nonzero even at a temperature of absolute zero. [69]

Adding particles to the system, or reducing its volume, forces the particles into still higher energy quantum states. This requires a compression force, which in turn produces a resisting pressure. The key properties are that despite degenerate matter still has a thermal pressure, the degeneracy pressure depends only on the density of the fermions, and it is this pressure that keeps the dense star in equilibrium, until the pressure, by the addition of mass, increases to the limit beyond which the degeneracy pressure cannot support the object against a collapse. When the collapse takes place, the maneuver room of the wave-particles is infinitely small, while their velocity will approach the speed of light, which results in a colossal pressure, and since the speed v of the wave-particles becomes so fast, we have to take relativity into consideration.

When the collapse occurs, we assume for symmetrical reasons that the star at the beginning collapses in an orderly manner. Let r be the distance between a wave-particle with mass m and the center of mass (CM), and let F'_G be the relativistic gravitational force that stems from the part of the spherical body, which lies within the radius r from CM. Finally, let v be the velocity of the wave-particle towards the center of mass, so the mass m moves directly against the gravitational center with the mass M.

Let us further assume that the center of mass is at rest relative to the ZPF, so the velocity v of the wave-particle towards CM, will also be its velocity relative to the ZPF.

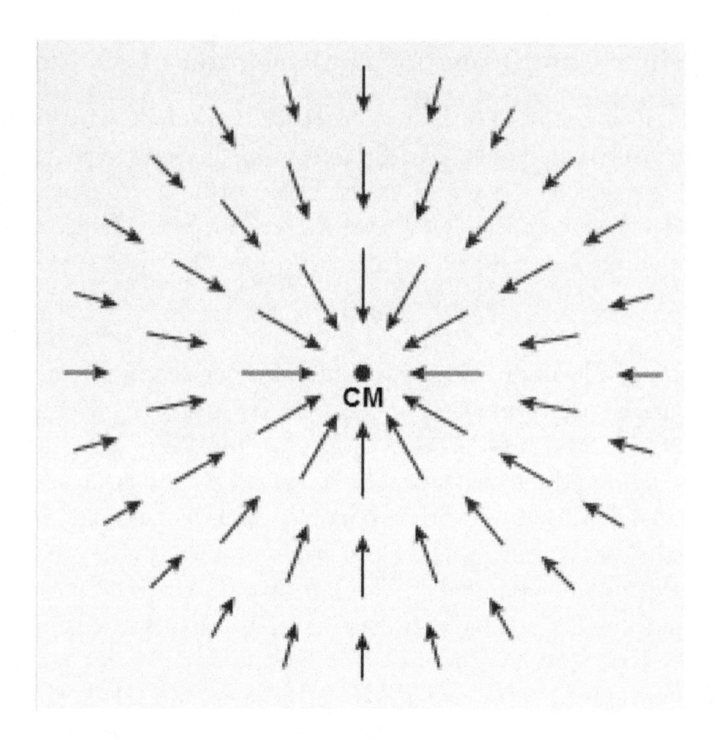

Figure 3.3: The collapse of a heavy star.

Since all the particles participate in the collapse, the number of particles within the radius r will be constant, which means, that we can assume that the force F'_G that stems from the mass M is constant on a sphere with radius r, and points in the direction of the center of mass.

From the expression for the relativistic gravitational force (eq. 2.56), we find the force between the center of mass with the mass M, and the mass m of the wave-particle:

$$F'_G = G\frac{mM}{r^2(1 - v^2/c^2)}.$$

The mass of the collapsing star within the radius r equals:

$$M = \sum_{i=1}^{n} m_i,$$

where all the wave-particles lie within the radius r, from the center of mass. From the equation it can clearly be seen that $F'_G \to \infty$, when $r \to 0$, and since:

$$F'_G = ma = G\frac{mM}{r^2(1 - v^2/c^2)} \Leftrightarrow a = \frac{GM}{r^2(1 - v^2/c^2)} \Leftrightarrow dv = \frac{GM}{r^2(1 - v^2/c^2)}dt, \quad (3.1)$$

it is seen that the velocity v of the wave-particles approaches the speed of light when the distance r approaches zero.

When the velocity v of the wave-particles approaches the speed of light, the force F'_G on the wave-particles will approach infinity. If the celestial object for instance is a white dwarf, the gravitational collapse will lead to an explosion where the collapsing star spews out vast amounts of material into the existing Universe.

3.2.3 Astronomical Considerations of Heavy Stars

A white dwarf is the final evolutionary state of a star with a mass less than around 8 solar masses, where nuclear fusion has powered the star for most of its life, and whose mass is not high enough to become a neutron star. This includes most of the stars in our galaxy. [70]

When a main-sequence star of low or medium mass runs out of hydrogen within its core, the balance between the outward pressure from the fusion of hydrogen into helium and the inward pressure of gravity tips, in the favor of gravity. However, the inward pull creates a pressure that causes the star to heat up again, which makes it able to fuse what little hydrogen that remains in a shell wrapped around its core.

The burning shell of hydrogen expands the outer layers of the star. When this happens, the star becomes a red giant. The core temperature of the red giant now increases until it is hot enough to fuse the helium that was created from the fusion of hydrogen. This will eventually transform the helium into carbon and other heavier elements by the triple-alpha process. [71]

Since the star will not be hot enough to ignite the carbon at its core, it will succumb to gravity once more. When the core of the star implodes, it will cause a release of energy that causes the envelope of the star to expand, so that the star becomes even larger than before. This continues until the star finally blows off its outer layers. However, the core of the star becomes a white dwarf. The white dwarf will be surrounded by an expanding shell of gas known as a planetary nebula, which is heated by the radiation from the white dwarf.

The mass of the white dwarf is very large, and equals the mass of the Sun, despite its volume only being slightly bigger than that of the Earth. Since no fusions take place in a white dwarf, its luminosity stems from the emission of stored thermal energy, and if it never achieves the necessary mass for an explosion, it will ultimately end up as an inert black dwarf. [71]

However, since the material in the white dwarf no longer undergoes any fusion, there is no obstacle for the pressure to place the nuclei closer to one another than normally allowed by the electron orbitals of normal matter. [72] So, if the white dwarf either by accumulation or merger achieves the necessary mass for an explosion, the pressure will rise inside the white dwarf, so the matter will be changed from normal matter composed of atoms joined by chemical bonds to a plasma of unbound nuclei and electrons.

When the plasma of a white dwarf is cooled and the energy to keep the atoms ionized is no longer sufficient to prevent a further collapse, the electrons must satisfy the Pauli exclusion principle, [73] which states that no two electrons can occupy the same quantum state. It means that at temperatures around zero all the electrons cannot occupy the lowest energy state, but will have to form a band of the lowest available states. This state of the electrons is called degenerate, but despite the temperature equals zero, the electrons still possess a high energy and pressure. [74] This means that the white dwarf has become a stellar core remnant that is primarily composed of electron degenerate matter, which is supported against further collapse by electron degeneracy pressure.

Since the pressure depends only on the density and not the temperature, a white dwarf can, by drawing on a reservoir of matter and energy in the vicinity, reach the Chandrasekhar limit, which is the limit beyond which the electron degeneracy pressure cannot support the star. The core will then collapse and explode as a supernova. This occurs when a white dwarf has reached a mass of approximately 1.4 $M_{\odot}$, [75] and can leave behind a neutron star. [73], [71], [76] However, other types of white dwarfs exist, such as, for instance, the carbon-oxygen white dwarf, which when it approaches the Chandrasekhar limit may explode as a type Ia supernova via a process known as carbon detonation. [73], [71]

Neutron stars are the densest stars known to exist. Besides arriving from a white dwarf, a neutron star can also result from a collapsed core of a large star, at between 8 and 29 $M_{\odot}$. [77] A neutron star is supported against further collapse by the neutron degeneracy pressure, and has a mass of between 1.1 $M_{\odot}$ [78] and 3.2 $M_{\odot}$. [79] The upper limit to the mass of stars composed of neutron degenerate matter is the Tolman–Oppenheimer–Volkoff limit, beyond which it cannot be supported by neutron degeneracy pressure, so a neutron star that approaches this limit will collapse into other forms of degenerate matter, such as quarks and gluons. [80]

3.3 Degenerate Matter

The Pauli exclusion principle does not allow two identical half-integer spin particles, or two fermions, to simultaneously occupy the same quantum state. The result is an emergent pressure against compression of matter into smaller volumes of space. For this reason, solid matter is stabilized by a quantum degeneracy pressure, where degeneracy

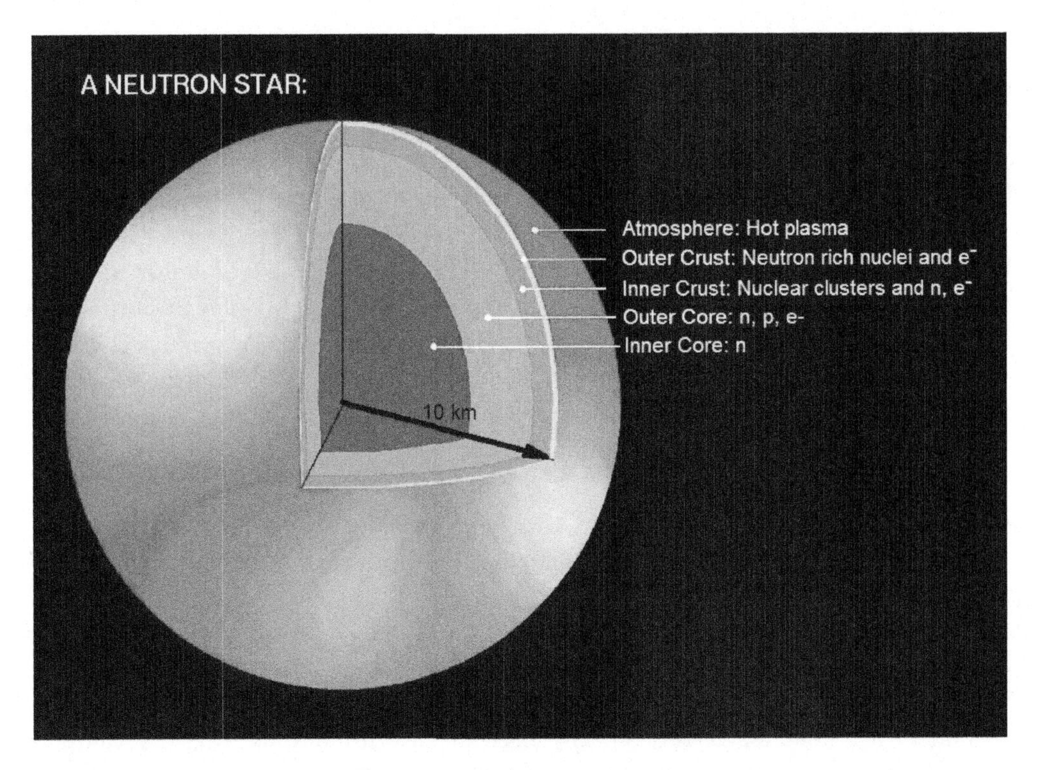

Figure 3.4: A neutron star.

creates a barrier against a gravitational collapse of dying stars, and is for instance responsible for the formation of white dwarfs and neutron stars. [81]

When fermions are squeezed too closely together, the exclusion principle requires that they have different energy levels. To add yet another fermion to a given volume, it requires that a fermion raises its energy level to make room, and this requirement for energy manifests itself as a pressure. This degeneracy pressure is omnipresent and is in addition to the normal gas pressure. At commonly encountered densities, this pressure is so low that it can be neglected. However, when the density is high enough, and the temperature is low enough, matter becomes degenerate, so the material is dominated by the degeneracy pressure. [82]

A useful way to explain the degeneracy pressure is the Heisenberg uncertainty principle, which states that:

$$\Delta x \Delta p \geq \frac{\hbar}{2}, \tag{3.2}$$

where Δx is the uncertainty in the position and Δp is the uncertainty in the momentum. A material subjected to an ever-increasing pressure will become more compact, and its delocalization, Δx, will decrease. Thus, as dictated by the uncertainty principle, the spread of the momenta of the fermions, Δp, will increase. So, no matter how low the temperature drops, the fermions must be traveling at this "Heisenberg speed", which contributes to the pressure. When the pressure due to this "Heisenberg motion" exceeds the pressure from the thermal motions of the fermions, the fermions are referred to as degenerate, and the material is termed as degenerate matter.

Electron degeneracy pressure will halt the gravitational collapse of a star, if its mass is below the Chandrasekhar limit of 1.39 solar masses. [75] Consequently, it is this degeneracy pressure that prevents a white dwarf star from collapsing. A star exceeding this limit, without having a significant thermally generated pressure, will continue to collapse to form a neutron star, because the degeneracy pressure provided by the electrons is weaker than the inward gravitational pull.

Degenerate gas, which is also called degenerate matter or Fermi gas, encompasses gases composed of fermions such as electrons and neutrons. [83] It is a collection of free, non-interacting particles, where the pressure and other physical characteristics are determined by quantum mechanical effects. Following the Pauli exclusion principle, there can only be one fermion occupying each quantum state, meaning that two particles cannot both have identical velocities, spins, and positions at the same time. The degenerate state of matter arises at an extraordinarily high density in compact stars. Examples of degenerate matter include electron degenerate matter and neutron degenerate matter. [84]

Normal gas exerts higher pressure when it is heated and expands, but the pressure of a degenerate gas does not depend on the temperature. When gas becomes super-compressed, the particles position right up against each other to produce a degenerate gas. Instead of temperature, the pressure in degenerate gas depends only on the density of the degenerate particles; however, adding heat to degenerate matter does not increase its density. [85] The mass density of a degenerate matter can only be increased through an increase of the number of particles, or through an increase of the external pressure, which raises the internal pressure, thus squeezing the particles closer together. This implies that their momenta are extremely high. So, even if the plasma is cold, the particles must be moving very fast on average. This leads to the conclusion that the particles under the extreme pressure have speeds around the speed of light at the time of the collapse. Such a phase of strong interacting neutrons is expected in the core of a massive neutron star.

3.3.1 The Neutron Degeneracy Pressure

Neutrons, which are composed of three quarks, are able to generate the same kind of degeneracy pressure as electrons, but first at densities of approximately 2.0×10^{18} kg/m^3. [86] This requires an amount of mass, so the radius of the core of the neutron star will be about 12.5 km. [87]

To evaluate the neutron degeneracy pressure we may once again apply the Heisenberg uncertainty principle,

$$\Delta x \Delta p \geq \frac{\hbar}{2},$$

where Δx is the uncertainty of the position, and Δp is the uncertainty of the momentum. During the accumulation of mass the neutron star can become a black hole on its way to a regenerative process. When the neutron star is exposed to an increasing pressure, it will compact more, so the displacement of the neutrons, Δx, will decrease, while the momenta of the neutrons, Δp, will increase, according to the uncertainty principle. No matter how low the temperature drops, the neutrons must be traveling at still higher speeds, which contributes to the pressure. Finally, when the pressure due to the degenerate motion exceeds the pressure from the thermal motions of the neutrons, the neutrons become degenerate, and the material is termed degenerate matter. [88] The neutron degeneracy pressure will halt the gravitational collapse of the star as long as its mass is below the Tolman-Oppenheimer-Volkoff limit. When the neutron degeneracy pressure exceeds this limit, the degeneracy pressure caves in, and gravity takes over, so the body collapses.

Although asymptotic freedom would appear to suggest otherwise, even at supranuclear densities (i.e. higher than that of a nucleus), nucleons cannot spontaneously be dissolved into their constituents of up and down quarks, as the overall energy per baryon would be higher. This might for instance be the case in the region inside a neutron star where the central density has reached the density of nuclear deconfinement. [89] A transition between the confined hadronic phase and the deconfined quark phase is triggered by the creation of a drop of the new stable phase. [90] This is a very common phenomenon in nature, as in fog or dew formation in supersaturated vapor, or ice formation in supercooled water. The astrophysical consequences of the creation of quark matter inside a massive hadronic compact star, such as a neutron star in which no fraction of quark matter is present, are that this first seed of quark matter will trigger the conversion from pure hadronic matter to quark matter, which is a first-order phase transition.

3.3.2 The QCD phase diagram

The ZPF in which we live, contains among the other fields the quantum chromodynamic (QCD) vacuum, which has the hadrons as its excitations. [91] However, when a neutron star or a black hole with a neutron star at its center is under an extremely high pressure from its own gravity, the hadrons are just one phase of quantum chromodynamics.

When the pressure in the dense interior of a neutron star reach the Tolman-Oppenheimer-Volkoff limit, the degenerate neutron star may collapse under the high pressure. Since the quarks are held together by the strong interaction, which is mediated by the gluons, the high baryon density phases can, because of the high density phases of QCD with

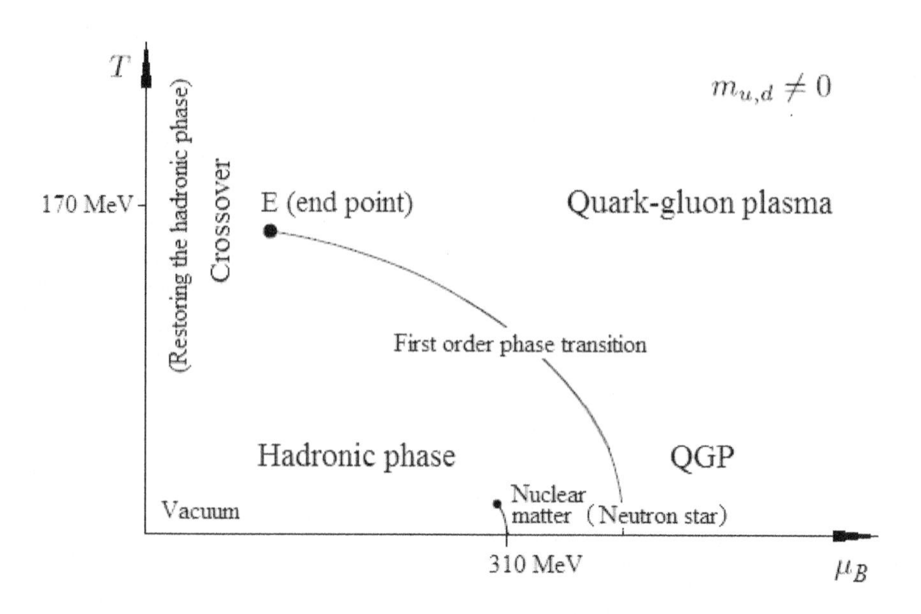

Figure 3.5: QCD phase diagram of the temperature T vs. the baryochemical potential μ_B.

their asymptotic freedom, be described in terms of quarks and gluons. [14], [91] As a result of asymptotic freedom the strong interaction becomes weaker when the pressure and velocity rise even more. Eventually, the color confinement would be lost and a plasma of freely moving quarks and gluons would be formed. This theoretical phase of matter is called quark–gluon plasma (QGP).

A QGP is a state of matter in quantum chromodynamics, which exists at extremely high temperatures and/or densities. Such a distinct state of deconfined quarks can be distinguished from any other state, and arises due to a first-order phase transition. [90] However, although asymptotic freedom would appear to suggest otherwise, even at supranuclear densities (i.e. higher than that of a neutron), neutrons cannot spontaneously dissolve into their constituent quarks, as the overall energy per baryon would be higher. But, if we consider the region inside a neutron star, where the central density has reached that of nuclear deconfinement, quantum fluctuations might initiate an abrupt change at the star's center, which could trigger a core-collapse of the neutron star, whose central density has increased due to spin-down or accretion. [89]

Such a conversion, where the individual neutrons break down into their constituent quarks of one up and two down quarks, might be confined to the center of a neutron star inside a black hole. [92] In the course of the conversion two processes take place: a gravitational collapse and a generation of energy. During the gravitational collapse the body implodes under the influence of its own gravitational force, which results in a body that is many times smaller and denser than the original celestial body, and which due to the collisions creates a plethora of exotic particles. Simultaneously, the binding energy of the gluons will, according to the mass-energy equivalence equation $E = mc^2$, release an abundance of energy, which is equal to the mass difference of between the neutrons and the resulting up quarks and down quarks.

The total energy liberated during the stellar conversion, E_{conv}, is given by the difference between the mass of the hadrons that takes part in the conversion, M_h, and the mass of the resulting quarks, M_q, so: [90], [93]

$$E_{conv} = (M_h - M_q) \cdot c^2. \tag{3.3}$$

The stellar conversion of a neutron star with a radius of about 12.5 km to its constituent quarks, [87] liberates an energy of the order of $E_{conv} = 10^{53}$ erg $= 10^{58}$ MeV. [93], [94] This huge amount of released energy will cause a gigantic explosion, which could be the energy source of a powerful gamma-ray burst or an explosion of a whole star. [94]

Let us now look at the phase diagram when the baryon number density is equal to or approaching zero ($\mu_B \sim 0$), with a brief review of the phase changes which occur as a function of temperature. [95] That is to say, we restrict ourselves to the vertical axis of the figure. This slice of the phase diagram is run through by the quark-gluon phase during the first tens of microseconds after a regenerative process, where the temperature just after the first-order phase transition is extremely hight. [91]

Since the quarks are not massless ($m_u = 2.0$ MeV/c^2 [96] and $m_d = 4.8$ MeV/c^2 [96]), physics changes dramatically but smoothly in the crossover region from the quark-gluon phase to the hadronic phase. So there is no sharp boundary on the vertical axis separating the high temperature quark-gluon plasma from the low temperature hadronic world. This picture is consistent with present lattice simulations, [97] which suggest that the energy of the quark-gluon plasma, at the point of the continuous transition from QGP to the hadronic phase, is roughly equal to $E \sim 140 - 170$ MeV. [98]

3.4 Neutron Stars

A neutron star can be formed by two separate processes. It can be produced when a medium-sized star ends up as a white dwarf or it can be produced from a star of about 8 solar masses. When a white dwarf by merger or accretion reaches the Chandrasekhar limit at about 1.4 solar masses, it will explode as a supernova, leaving behind a neutron star. [73], [71], [76] However, a neutron star can also be produced when any main sequence star, with an initial mass of above 8 solar masses, evolves away from the main sequence and end up as an iron-rich core that, by subsequent nuclear burning, creates a neutron star. In both cases the neutron star will have a radius of approximately 10 km and a mass between 1.1 and 3.2 solar masses that consists almost entirely of neutrons. [77],

[99] And when all the nuclear fuel in the core has been exhausted, the core will only be supported by the electron degeneracy pressure. Such neutron stars are the source of the largest explosions in the universe.

Let us take a closer look at the creation of a neutron star from a star with a mass above 8 solar masses, when the star has produced an iron-rich core. Until this stage the enormous mass of the star has been supported against gravity by the release of energy during the fusion of lighter elements into heavier ones. However, the formation of iron in the core terminates the fusion process, since iron is the most stable element. When further matter is added to the neutron star the temperature rises at the core, and when it reaches a temperatures of about 5×10^7 K, [100] photo-disintegration occurs, which is the breaking up of iron nuclei into alpha particles by high-energy gamma rays. When all nuclear fuel in the core has been exhausted, the core must be supported by degeneracy pressure alone. Further deposits of material from accretion or shell burning - where shell burning is the fusion of hydrogen, helium and other elements in the regions outside the stellar core - cause the core to exceed the Chandrasekhar limit. [101]

When the electron degeneracy pressure is overcome, the core collapses, sending temperatures soaring to about 10^9 K. [102] When such a star explodes as a supernova, the explosion is followed by a gravitational collapse, which compresses the core of the star to the density of the nucleus of an atom. During the collapse of the core, the rotation rate increases as a result of the conservation of the moment of inertia, hence the core can rotate up to several hundred times per second.

As the temperature climbs even higher, electrons and protons combine to form neutrons via electron capture, releasing a flood of neutrinos. When the densities reach the maximum stellar central density in the range of 4-8 times the saturation density of 2.8×10^{17} kg/m^3, [90] the neutron degeneracy pressure halts the contraction. The in-falling outer envelope of the star is halted and flung outwards by a flux of neutrinos produced during the creation of neutrons, the star has become a supernova. The remnant left behind is a neutron star - and if it has a mass greater than about 3 $M_{\odot}$, it becomes a black hole. [17]

The atmosphere of a bare neutron star is expected to be chemically very pure and dominated by hydrogen and helium, [103] whose dynamics is fully controlled by the neutron star's magnetic field, [104] which can reach a strength of up to 10^{11} tesla on the surface of a neutron star. [105] At such magnetic field strengths the neutron stars are named magnetars. The bare neutron star's gravity accelerates the in-falling matter to tremendous speeds. The force of its impact would likely destroy the object's component atoms. Below the atmosphere there must be an extremely hard and very smooth crust, with maximum surface irregularities of about 5 mm, due to the extreme gravitational field. [106]

The nuclei at the solid surface are probably iron, because of iron's high binding energy per nucleon. [103] If the surface temperature exceeds 10^6 Kelvin, as in the case of a young pulsar, the surface should be fluid, instead of the solid phase that might exist in cooler neutron stars. [103] The temperature inside a newly formed neutron star is from around 10^{11} to 10^{12} Kelvin. [107] However, the huge number of neutrinos it emits carry away so much energy that the temperature of an isolated neutron star falls within a few years to around 10^6 Kelvin. [107]

Proceeding inward, nuclei are encountered with ever-increasing numbers of neutrons. Such nuclei would decay quickly on Earth, but inside the neutron star the nuclei are kept stable by the tremendous pressures. As the process of photo-disintegration and the formation of neutrons via electron capture continues at increasing depths, the content of neutrons becomes overwhelming. In that inner region, there are nuclei, free electrons, and free neutrons. Furthermore, the nuclei become increasingly smaller as gravity and pressure overwhelm the strong force, until the core is reached, where the composition of the superdense matter can be described as neutron-degenerate matter, composed of neutrons and perhaps some protons and electrons. [108]

The neutron star's density varies from about 1×10^9 kg/m^3 in the crust, increasing with depth to between 6×10^{17} and 8×10^{17} kg/m^3 deeper inside, [107] which is denser than an atomic nucleus. So neutron stars have overall densities of 3.7×10^{17} to 5.9×10^{17} kg/m^3, and the pressure increases from 3.2×10^{31} to 1.6×10^{34} Pa from the inner crust to the center. [109] The gravitational field at a neutron star's surface is about 2×10^{11} times stronger than on Earth, at around 1.86×10^{12} m/s^2. [110] Such a strong gravitational field acts as a gravitational lens and bends the radiation emitted by the neutron star, so parts of the normally invisible rear surface become visible (see eq. 2.55). [111] The gravitational force of a typical neutron star is so huge that if an object were to fall from a height of one meter on a neutron star 12 kilometers in radius it would reach the ground with a velocity of around 1.4 million meters per second, or 5 million kilometers per hour. [112]

If further mass is added to the neutron star, the star may collapse when it exceeds the Tolman–Oppenheimer–Volkoff limit of 3.2 solar masses $\sim 6.4 \times 10^{30}$ kg, which is thus the maximum mass of a pure neutron star. [79]

3.4.1 The Speed of the Neutrons inside a Neutron Star

When the innermost part of the neutron star reaches the Tolman–Oppenheimer–Volkoff limit, the velocity of the neutrons approaches the speed of light, which can be seen from the Heisenberg uncertainty principle, $\Delta p \Delta x \geq \hbar/2$, where Δp is the uncertainty in the particle's momentum and Δx is the uncertainty in position. The uncertainty principle contains implications about the amount of energy which is required to contain the particles within a given volume, and Planck's constant determines the size of the confinement that can be produced by these forces. Since the star is under an extreme pressure, the neutrons must be located in a very confined space.

The neutron, which consists of two down quarks and one up quark, has a root mean square radius of about: $r_n = 0.8 \times 10^{-15} = 0.8$ fm, [113] which is composed of a positively charged core of radius ≈ 0.3 fm surrounded by a compensating negative charge with a radius between 0.3 fm and 2.0 fm. [113] So, the root mean square diameter of the neutron is $d_n = 1.6$ fm, while the neutron diameter ranges from 1.2 fm to 4.6 fm.

The following calculation serves to give an estimation of the velocities of the neutrons, when they are squeezed together within a neutron star, such that the diameter d_n of the neutron equals their physical extent Δx. According to the uncertainty principle we have $\Delta x \Delta p \geq \hbar/2$, where $\hbar \approx 1.05 \times 10^{-34}$ kg·m^2/s. For the minimum value of the diameter we find the following value of the momentum, Δp:

$$\Delta p \geq \frac{\hbar}{2\Delta x} = \frac{1.05 \times 10^{-34}}{2 \cdot 1.2 \times 10^{-15}} \text{ kg·m/s} = 4.4 \times 10^{-20} \text{ kg·m/s}.$$

If we set $\Delta p = p$ the internal energy of the neutron will be equal to:

$$E = \frac{mv^2}{2} = \frac{p^2}{2m},$$

from which we find the velocity of the neutron:

$$v = \sqrt{\frac{p^2}{m^2}}.$$

Since the mass of a neutron is equal to $m_n = 1.7 \times 10^{-27}$ kg, [114] the velocity v_n corresponds to:

$$v_n = \sqrt{\frac{p^2}{m^2}} = \sqrt{\frac{(4.4 \times 10^{-20})^2}{(1.7 \times 10^{-27})^2}} \text{ m/s} \approx 26.000 \text{ km/s}.$$

This is the velocity of the neutrons when they are placed next to each other. However, during the development towards an explosion, the neutron star will need to accumulate more mass. When it reaches a mass that is greater than about 3 $M_\odot$, [17] the neutron star becomes a black hole. Since a bare neutron star needs a mass equal to 3.2 $M_\odot$ in order to exceed the Tolman–Oppenheimer–Volkoff limit, it will become a black hole just before the explosion. At that time the neutrons potentially have a velocity that approaches the speed of light, despite the neutrons at the time of the explosion are squeezed tightly together. [115]

3.4.2 The Ratio between the Volume of the Neutron and the Quarks

To evaluate how much room there is for such a collapse, it is essential to know the proportion between the volume of the particles before and after the collapse. The particles before the collapse consist almost entirely of neutrons, while the result of the collapse is two down quarks, and an up quark.

When we use the root mean square radius of the neutron, $r_n = 0.8 \times 10^{-15} = 0.8$ fm, [113] the calculated volume of a neutron, V_n, will be equal to or greater than the actual volume of a neutron that is placed inside a collapsing neutron star:

$$V_n = 4\pi r_n^3/3 = 4 \cdot \pi \cdot (0.8 \times 10^{-15})^3/3 \text{ m}^3 = 1.1 \times 10^{-46} \text{ m}^3.$$

The size of the quarks is not exactly known, but from experimental data it is possible to estimate an upper bound of the quark's radius. The experimental data originate from measurements of the normalized dijet angular distributions of high energy proton-proton collisions. The distributions are in agreement with predictions of quantum chromodynamics, which set the lower limit on the contact interaction to 7.5 TeV. [116]

According to the Planck-Einstein relation we know that $\lambda = hc_0/E$ for radiation with the speed of light. The relation can be derived from Planck's radiation law $E = hf$ (eq. 2.15) and the connection between the frequency, wavelength and the speed of light: $c_0 = f\lambda$ (eq. 2.18), so:

$$E = hf = hc_0/\lambda.$$

The following calculation is just a rough estimation, but it is as close as we can get to declaring an upper bound on the quark's radius, with the knowledge we have today. The radius is thus:

$$r_q = \frac{\lambda}{2\pi} = \frac{h \cdot c_0}{2\pi \cdot E} = \frac{6.6 \times 10^{-34} \cdot 3.0 \times 10^{8}}{2\pi \cdot 7.5 \times 10^{12} \cdot 1.6 \times 10^{-19}} \text{ m} = 2.6 \times 10^{-20} \text{ m}.$$

Since the radius of a quark is $r_q \leq 2.6 \times 10^{-20}$ m, the volume of a quark is approximately equal to:

$$V_q = 4\pi r_q^3/3 \leq 4 \cdot \pi \cdot (2.6 \times 10^{-20})^3/3 \text{ m}^3 = 7.4 \times 10^{-59} \text{ m}^3.$$

That is, the proportion between the volume of the neutron and the three quarks is about $V_n/3V_q = 1.1 \times 10^{-46}/(3 \cdot 7.4 \times 10^{-59}) = 5.0 \times 10^{11}$, so there is plenty of room for a collapse of a neutron star.

3.4.3 The Release of Energy when a Neutron is Converted to Quarks

To make an estimation of how much energy there is released when the neutrons are converted into quarks, it is suitable to calculate the difference between the mass of the neutrons and the mass of the two down quarks and the up quark; and since force-carrying particles, such as virtual photons, gravitons, and gluons, do not have any mass, [117] we do not need to take these particles into consideration.

The mass of the neutron: $m_n = 1.67 \times 10^{-27}$ kg. [114]
The mass of a down quark from (lattice calculations): $m_d = 4.79$ MeV/c^2. [118]
The mass of an up quark (from lattice calculations): $m_u = 2.01$ MeV/c^2. [118]

So, the total mass (m_q) of the two down quarks and the up quark, which make up the neutron, is:

$$m_q = 2 \cdot 4.79 \text{ MeV/c}^2 + 2.01 \text{ MeV/c}^2 = 11.59 \text{ MeV/c}^2,$$

or since 1 eV $\simeq 1.60 \times 10^{-19}$ J, and the speed of light corresponds to 3.0×10^8 m/s, the mass of the quarks in kilograms is:

$$11.59 \text{ MeV/c}^2 = 11.59 \times 10^6 \cdot 1.60 \times 10^{-19}/(3.0 \times 10^8)^2 \text{ kg} = 2.06 \times 10^{-29} \text{ kg}.$$

The difference between the mass of the neutron and the mass of the quarks that make up the neutron is then:

$$\Delta m = 1.67 \times 10^{-27} kg - 2.06 \times 10^{-29} \text{ kg} = 1.65 \times 10^{-27} \text{ kg}$$

which means that each neutron releases an energy equal to:

$$E_n = \Delta m \cdot c^2 = 1.65 \times 10^{-27} \cdot (3.0 \times 10^8)^2 \text{ J} = 1.49 \times 10^{-10} \text{ J or } 928 \text{ MeV}.$$

For comparison the total energy production of one whole chain reaction in the Sun, where four protons are converted into a helium-4 atom, amounts to 26.73 MeV. [119]

3.4.4 The Energy Content of the Core of a Neutron Star

From the Tolman-Oppenheimer-Volkoff limit we know that there is an upper limit to the density, and thereby to the mass and size of a neutron star. There is, however, a great uncertainty regarding the size of the Tolman-Oppenheimer-Volkoff limit. This uncertainty in the value reflects the fact that the equations of state for extremely dense matter are not well known. However, some physicists set the upper size of a neutron star equal to 3.2 $M_\odot$, [79] according to which the maximum mass of a neutron star (NS) is:

$$M_{NS} = 3.2M_{\odot} = 3.2 \cdot 1.99 \times 10^{30} \text{ kg} = 6.37 \times 10^{30} \text{ kg},$$

where, $M_{\odot} = 1.99 \times 10^{30}$ kg, [120] is the solar mass.

Since each neutron releases the energy $E_n = 1.49 \times 10^{-10}$ J and the mass of the neutron is $m_n = 1.67 \times 10^{-27}$ kg, [114] the total energy content of a neutron star must be equal to:

$$E(\text{NS}) = \frac{6.37 \times 10^{30}}{1.67 \times 10^{-27}} \cdot 1.49 \times 10^{-10} \text{ J} = 5.7 \times 10^{47} \text{ J} \simeq 3.6 \times 10^{60} \text{ MeV},$$

as 1 J $\simeq 6.242 \times 10^{12}$ MeV.

Some researchers set the density of the neutron star at the time of the explosion to at least 1.0×10^{18} kg/m^3. [86] So a cubic meter of such a neutron star must at least release the energy:

$$E(1m^3 \text{ NS}) = \frac{1.0 \times 10^{18}}{1.67 \times 10^{-27}} \cdot 1.49 \times 10^{-10} \text{ J/m}^3 = 8.92 \times 10^{34} \text{ J/m}^3.$$

Since the radius of a neutron star gets smaller, when the density increases, the maximum radius of a neutron star will be around 11.5 km at the time of explosion. [121] So, the energy content will at least be equal to:

$$E(\text{Explosive NS}) = 8.92 \times 10^{34} \cdot 4\pi \cdot (11.5 \times 10^3)^3/3 \text{ J} = 5.68 \times 10^{47} \text{ J} \simeq 3.5 \times 10^{60} \text{ MeV}.$$

This is the amount of energy that is generated, when a neutron star reaches the Tolman-Oppenheimer-Volkoff limit. The explosion begins when the innermost part of the neutron star has reached the TOV limit, and a seed of quark matter triggers the conversion from hadronic matter to quark matter. This initial explosion creates a shock wave, which, due to the continued generation of energy propagates outward through the star, generating an explosion of the neutron star.

3.4.5 A Black Hole with a Neutron Star at its Center

Many of the black holes at the center of the galaxies have created super-massive black holes (SMBHs) by devouring the substance in the vicinity, so there is a super-massive black hole in the center of virtually all the galaxies. Since the mass of a black hole is at least 3 $M_{\odot}$, [17] and the mass of a neutron star is between 1.1 and 3.2 $M_{\odot}$, [79] all the massive black holes at the center of the galaxies can be expected to contain a neutron star at the center. SMBHs with masses equivalent to billions of solar masses have been found, [65] and most of the SMBHs act as AGNs, some of which are quasars or other forms of active galactic nuclei.

The gravitational field at the surface of a black hole with a neutron star at its center is very high and accelerates the incident matter to extreme speeds. Determined by the nature of the added material, fusion of hydrogen, helium, carbon, oxygen, silicon and iron can occur on the outer shells of a black hole, producing still heavier atomic nuclei until the fusion of iron. As the pressure rises within the black hole, temperatures rise to over 5×10^9 K. [122] At these temperatures, photo-disintegration occurs, which is the breaking up of iron nuclei into alpha particles by high-energy gamma rays. Closer to the center of the black hole the temperature climbs even higher, so electrons and protons combine to form neutrons via electron capture, releasing a flood of neutrinos. When the gravitational pressure overwhelms the strong force, the neutrons become a collection of free, non-interacting neutron-degenerate matter, [83] called neutronium, where the extreme pressure can deform the neutrons into a cubic symmetry, allowing a tighter packing of the neutrons at the center of the black hole. [123]

The pressure at the core of a black hole, with a neutron star at its center, is now so high that an explosion is imminent. However, there are still some parameters that may influence the course of the eruption. The rotational speed and the internal structure of the black hole may have an influence on the time and nature of the explosion. When further material is added to the surface of the black hole, the pressure will rise at the interior, and depending on the distance from the center, different processes in the different shells surrounding the core occur, where the structure of the shells chages from the newly arrived material at the surface of the black hole to the neutron degenerate matter at the center. Since the final stage of fusion is the production of iron during silicon burning, the giant black holes can form shells of different materials, which for a time prevent the build-up of a sufficient pressure to create an explosion.

When the neutron star, as a result of the increasing external pressure from the black hole, finally reaches the Tolman-Oppenheimer-Volkoff limit, the neutrons break down into their constituent quarks and gluons, [124] creating a regenerative process during which the shells are wrecked. The combination of the direction and speed of the rotation, the strength and direction of the magnetic field, the size of the black hole, the constitution of the shells and the rate of accretion of matter are all parameters which have an influence on the final course of the explosion. SMBHs can be anything from dormant to providing a continuous supply of energy to the neutron star, which in turn can supply the energy for e.g. a quasar.

3.4.6 The Relation between the Size of a Neutron Star and a Black Hole

Observations of distant luminous quasars show that there are SMBHs with masses corresponding to billions of solar masses as far away as we can see. The black holes with a neutron star at the center are the best candidates in the Universe for the creation of AGNs, where densely packed cold neutrons at the center of a neutron star inside a black hole exceed the Tolman–Oppenheimer–Volkoff limit due to the immense external pressure. This creates an explosive first order phase transition from densely packed neutrons to a phase of strong interacting deconfined quarks and gluons during the release of a huge amount of energy. It is seen that the densely packed neutrons fulfill the conditions of a low temperature T, and a high baryon chemical potential μ as illustrated in the phase diagram of chromodynamics (Fig. 3.5), [90] where the interior composition of the neutron star at the Tolman–Oppenheimer–Volkoff limit is pure hadronic matter before the explosion and a mixed phase of quarks, gluons and energy after the explosion, [125] which immediately creates a galore of particles and electromagnetic radiation.

A neutron star inside a black hole, which is a candidate for an explosion, can start at a radius of about 12.5 km; [87] however, as the external pressure on the black hole with the neutron star at the center increases due to accretion of matter from the surroundings, the radius of the neutron star R_{NS} decreases. When the density of the densely packed neutrons rises to 6 times the saturation density, the radius of the neutron star will shrink to about 9.7 km, where the neutron star reaches the Tolman–Oppenheimer–Volkoff limit. At this time the neutron star has a mass of around 3.2 $M_\odot$ where $M_\odot = 1.99 \times 10^{30}$ kg.

In this way, neutron stars represent the limit where gravity overcomes all the other forces in nature. In fact, the structure of neutron stars based on the equations of state of hadronic matter predicts a maximum stellar central density in the range of 4-8 times the saturation density $\rho \sim 2.8 \times 10^{17}$ kg/m^3 of nuclear matter, [126] which is comparable to the approximate density of an atomic nucleus of 3×10^{17} kg/m^3. This entails that a dense and cold neutron phase can arise at the center, so the core of a neutron star inside a black hole is the best candidates in the Universe where a phase of strongly interacting matter of deconfined quarks and gluons can arise.

Let us consider an idealized situation with a non-rotating spherical neutron star. Since it is reasonable to think of a neutron star under gravitational pressure as a fluid, where the density is greatest at the center, we can choose the average stellar density to be about 6 times the saturation density. The volume of a neutron star V_{NS} with a mass equal to 3.2 $M_\odot$ and an average saturation density equal to $6\rho = 6 \cdot 2.8 \times 10^{17}$ kg/m^3 is then:

$$V_{NS} = \frac{3.2 M_\odot}{6\rho},$$

where $M_\odot = 1.99 \times 10^{30}$ kg. Since the volume of a sphere is equal to $V = 4\pi R^3/3$, the radius R_{NS} of a neutron star at the time of the explosion can be expressed as:

$$R_{NS} = \sqrt[3]{\frac{3 \cdot V_{NS}}{4\pi}} = \sqrt[3]{\frac{3 \cdot 3.2 M_\odot}{4\pi \cdot 6\rho}} = \sqrt[3]{\frac{3 \cdot 3.2 \cdot 1.99 \times 10^{30}}{4\pi \cdot 6 \cdot 2.8 \times 10^{17}}} \approx 9.7 \text{ km}.$$

Since the neutron star cannot achieve the necessary density for an explosion without an external pressure, the outermost layers of the star must grow through accretion of matter from the surroundings; [127] however, due to the structure and composition of the outer layers and the rotational speed of the star, the candidates for the generation of regenerative processes can be very large, such as for instance the SMBHs at the center of the galaxies.

When the total mass reaches a size where the Schwarzchild radius becomes greater than the radius of the neutron star R_{NS}, such that: $r_{Schwarzschild} > R_{NS}$, the star becomes a black hole with a neutron star at the center.

If n is the number of solar masses, the total mass of the star can be expressed by: $n \cdot M_\odot$, so the neutron star becomes a black hole with a neutron star at the center, when:

$$r_{Schwarzschild} = \frac{2G \cdot n \cdot M_\odot}{c_0^2} > 9.7 \times 10^3 \text{ m},$$

where $c_0 = 3.00 \times 10^8$ m/s is the speed of light, $G = 6.67 \times 10^{-11}$ m^3 kg^{-1} s^{-2} is the gravitational constant, and $M_\odot = 1.99 \times 10^{30}$ kg is the solar mass. The number of solar masses must then fulfill the inequality:

$$n > \frac{9.7 \times 10^3 \cdot c_0^2}{2G \cdot M_\odot}$$

from which we find the number of solar masses needed to create a black hole with a neutron star at the center:

$$n > \frac{9.7 \times 10^3 \cdot (3.00 \times 10^8)^2}{2 \cdot 6.67 \times 10^{-11} \cdot 1.99 \times 10^{30}} = 3.3.$$

So the neutron star becomes a black hole with a neutron star at the center, when the total mass is greater than $3.3 \cdot M_\odot \approx 6.6 \times 10^{30}$ kg.

Since the neutron stars are dependent on an external pressure to generate the required density for an explosion, the brightest AGNs are seen at the center of the galaxies, where the neutron star resides inside a huge black hole. The explosion begins at the center of the neutron star, inside the black hole, where the internal pressure from the explosion transfers the particles to the surface of the black hole. Such black holes can have masses up to millions of solar masses, but the overall density of the huge black holes is by definition only just so large that the gravitational force at the surface of the black holes can hold on to the electromagnetic radiation. Since the free quark-gluon plasma can reach velocities up to $\sqrt{3}c_0$ (chapter 3.4.7), the quark-gluon plasma will be able to leave the black holes.

3.4.7 How Quark-Gluon Plasma can escape a Black Hole

Since the quarks in a quark gluon plasma are not bound together by forces that propagate at the finite speed of light relative to the ZPF, the quarks can be considered as free wave-particles whose velocities c are determined by the product of the wavelengths and the frequencies of the wave-particles, so their velocities are equal to $c = \lambda \cdot f$ (eq. 2.16).

According to equation 2.40, there is the following connection between the velocity c and the frequency f of a wave-particle located in a gravitational field at the distance r from the center of mass of a body, with the mass M:

$$f = f_0/[1 - c^2/(2c_0^2) + GM/(rc_0^2)].$$

Moreover, there is the following connection between the velocity c and the frequency f of a free wave-particle that is not affected by a gravitational field:

$$f = f_0/[1 - c^2/(2c_0^2)],$$

which means that a free wave-particle with the velocity c relative to the ZPF, which is not affected by a gravitational field, can obtain a velocity $c < \sqrt{2c_0^2} = \sqrt{2}c_0$, before its frequency becomes infinite.

Since the frequency according to the expression $f = f_0/[GM/(rc_0^2)]$ is dependent on the strength of the gravitational field, we will examine what influence the gravitational field from a neutron star has on the frequency, and thereby on the maximum velocities of the particles, which is crucial to their escape from a neutron star.

Due to the enormous gravity, the interior of the neutron star can be considered as a homogeneous liquid with an almost linear rise of the gravitational force, from zero at the center to the maximum at the surface of the neutron star. However, outside the surface, gravity decreases exponentially with the distance from the neutron star. (See [39] p. 377).

As the wave-particles experience the greatest force at the surface of the neutron star, we choose to look at the influence of gravity on the frequency of the wave-particles at the surface of the neutron star just before the explosion. At this time, the expression $GM/(rc_0^2)$ can be considered as a constant where the radius of the neutron star r equals $R_{NS} = 9.7$ km and the mass M corresponds to the maximum mass of the neutron star

$M_{NS} = 3.2M_{\odot} = 3.2 \cdot 1.99 \times 10^{30} = 6.37 \times 10^{30}$ kg, from which we get:

$$GM/(rc_0^2) = GM_{NS}/(R_{NS}c_0^2),$$

where the gravitational constant $G = 6.67 \times 10^{-11}$ m^3 kg^{-1} s^{-2} and the speed of light $c_0 = 3.00 \times 10^8$ m/s. So:

$$GM/(rc_0^2) = GM_{NS}/(R_{NS}c_0^2) = \frac{6.67 \times 10^{-11} \text{ m}^3 \text{ kg}^{-1} \text{ s}^{-2} \cdot 6.37 \times 10^{30} \text{ kg}}{9.7 \times 10^3 \text{ m} \cdot (3.00 \times 10^8 \text{ ms}^{-1})^2} = 0.5.$$

From the denominator in the expression: $f = f_0/[1 - c^2/(2c_0^2) + GM/(rc_0^2)]$, where we make use of the found value for $GM/(rc_0^2)$, it can be seen that the frequency approaches infinity when the denominator approaches zero, which is when:

$$1 - c^2/(2c_0^2) + GM/(rc_0^2) = 1 - c^2/(2c_0^2) + 0.5 = 0 \Leftrightarrow c^2 = 3c_0^2 \Leftrightarrow c = \sqrt{3}c_0.$$

Therefore, we can conhclude that it is physically possible for a free wave-particle to move at any velocity c less than $\sqrt{2}c_0$, and in the extreme cases where the wave-particle is under the influence of a gravitational field from a neutron star, it can move at any velocity less than $\sqrt{3}c_0$. Such velocities will, along with the high pressure, and the explosive expansion, make it possible for a hot quark-gluon plasma to escape a black hole.

We can further establish, that since the wave-particles get red-shifted when they leave the gravitational field of a black hole, they are able to achieve a larger velocity before the frequency becomes infinite.

3.5 Regenerative Processes

A regenerative process is defined as a process that converts heavy nuclei into lighter ones, such as hydrogen and helium, or, in other words, converts elements with a high atomic number, Z, into elements with a low atomic number and distributes them in the surrounding universe. This distribution of nucleons and electrons feeds the interstellar clouds from which the stars and galaxies are created. That such a conversion takes place is most obvious from the cosmic radiation that primarily consists of protons and alpha particles.

The largest observed regenerative explosions in the Universe originate from AGNs such as Seyferts and quarars which emit most of the quark-gluon plasma (QGP) along with the electromagnetic radiation. It is however not all eruptions that fall under the category of regenerative processes. For instance, supernovae fuse and eject a great deal of complex chemical elements, and thereby play a significant role in enriching the interstellar medium with heavier elements, but they are not known to create any lighter elements.

Since the quarks are not composite particles [128] and thereby not bound together by any forces that can deliver the energy for an explosion, and since neutron stars are the densest stars known to exist, it is reasonable to assume that the Seyferts, quasars and other AGNs are generated by black holes with a neutron star at the center. The eruptions may then take place, when the neutron stars along with the black holes, through accretion of matter, are able to generate a gravitational pressure, which is so huge that the density of the neutron stars exceeds the Tolman–Oppenheimer–Volkoff limit.

3.5.1 The Creation of Regenerative Processes

As mentioned, a regenerative process is a series of nuclear processes that converts heavy elements into lighter elements, such as hydrogen and helium, and is generally produced by a neutron star inside a black hole, at the center of a galaxy, where the black hole is surrounded by an orbiting accretion disk. In this way, the neutron star receives an outer shell of material that contributes to the internal pressure that generates the quark-gluon plasma.

From the proportion between the mass of a neutron and the mass of the three individual quarks that it is composed of, it can be seen that most of the mass of a neutron is the result of the strong force field, while the individual quarks only provide about one percent of the mass. When the first-order phase transition takes place, a greater or lesser part of the neutron star may take part in the process, [129] where the neutrons are converted into a hot, dense quark-gluon plasma. The strong force, which is carried by the gluons and binds the quarks together to form the neutrons, has such a high strength, that the strong force produces new particles, when the quarks split up at the Hagedorn temperature of 2×10^{12} K; [130] and the quarks that are not bound together by gluons, may act as free particles that can be heated above the Hagedorn temperature. [131]

When the temperature rises above the Hagedorn temperature, [132] asymptotic freedom will govern the quark-gluon plasma. This is analogous to conventional plasma, where nuclei and electrons can move freely around. However, the strength of the color force means that unlike the gas-like plasma, quark–gluon plasma behaves as a near-ideal Fermi liquid, [133] which flows with almost no friction. [134]

Since the force-carrying gluons themselves possess color charges, the gluon field between a pair of color charges forms a narrow flux tube between them. Because of this behavior, the strong force between the particles is constant regardless of their separation. [135] So, when two color charges are separated, it becomes, at some point, more energy-efficient for a new quark–antiquark pair to appear, rather than extending the tube further. As a result of this, jets of many color-neutral particles such as mesons and baryons are produced, instead of producing individual quarks. This process is called hadronization, where the intermediate mesons quickly decay into pairs of gamma-rays. [136]

After the plasma has left the black hole, it has the potential to expand during a transformation into a jet of photons, [137] nucleons, and leptons. [138] The distribution of nuclear material can be continuous or occur as relatively short bursts. As the particles are not bound together by forces which move with the finite speed of light relative to the ZPF, the particles can move with velocities up to between $\sqrt{2} \cdot c$ and $\sqrt{3} \cdot c$ in the central gravitational field (see chapter 3.4.7), which means that the plasma will be able to leave the black hole. When the neutron star rotates, the emissions are often concentrated around the direction of the magnetic field; however, in some rare cases the explosion can spread its material in all directions.

In the first fraction of a second after the material has left the celestial body, the eruption will consist of a hot quark-gluon plasma, which will almost immediately be converted to a mixture of free quarks, leptons and photons. In this initial jet the temperature will exceed 10^{12} K, and most of the material will consist of radiant energy. As the expansion continues during a continuous cooling, the radiation will decrease while other physical processes take over. After about a microsecond, depending on the pressure, the temperature may drop so much that the free quarks and gluons assemble into protons and neutrons. [139]

After several minutes, the temperature falls to around 10^9 K. This will allow for the formation of basic atomic nuclei such as protons, deuterons, and alpha particles. During the further expansion, the time between the nuclear collisions will rise, and the ratio between deuterons and alpha particles will stabilize. [140] At about 3000 K, the temperature is so low that the electrons can be captured by the atomic nuclei, and atoms can be formed.

The protons will most likely end up as cosmic rays or hydrogen in the interstellar medium, while the electromagnetic radiation, if it is not stopped, may reach the farthest regions of the closed Universe before the radiation gradually turns around as a part of the CMB.

Since the largest galaxies have the biggest black holes and the most violent explosions, the galaxies are self-regulating in the sense that the largest galaxies expel the most material, and depending on the size of the neutron stars (inside the black holes), their rotational speeds and the strength of their magnetic fields, the explosions can take on many different forms, such as quasars, magnetars, soft gamma repeaters (SGRs), and anomalous X-ray pulsars (AXPs).

3.5.2 A Black Hole's size is related to an AGN's Luminosity

It has been found that the relation between the mass of a central black hole, M_{bh}, and the stellar mass of the surrounding bulge, M_{bulge}, can be expressed by: [141]

$$M_{bh} \sim (1.2 \pm 0.06) M_{bulge}, \tag{3.4}$$

which means that there is a close relationship between the mass of a black hole, and the stars that comprise the elliptical galaxy or the central bulge of a spiral galaxy.

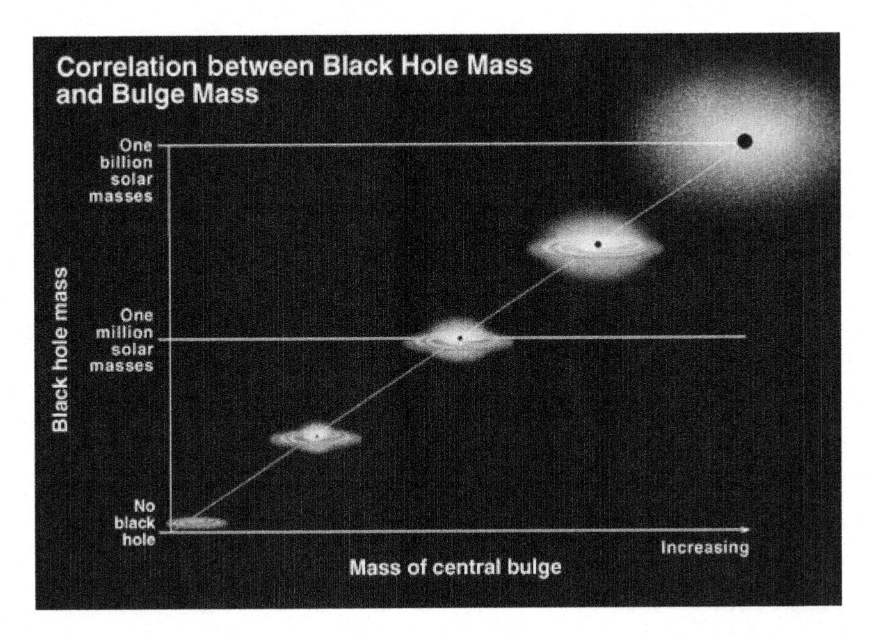

Figure 3.6: Correlation between the mass of a black hole and the bulge mass/brightness. [142]

It has also been found that there is a relation between the total energy E and the total mass M of the galaxies, and the total energy E and the total mass M of the clusters of galaxies, such that:

$$E^2/(GM^3) \equiv g, \text{ [143]} \tag{3.5}$$

where g is a constant, and G is the universal gravitational constant. The constant g has the dimension of an acceleration, and its value is $(5.6 \pm 0.9) \times 10^{-12}$ m/s for spiral galaxies, $(5.6 \pm 0.4) \times 10^{-11}$ m/s for ellipticals, and $(1.07 \pm 0.15) \times 10^{-11}$ m/s for clusters of galaxies. [143], [144] From the relation, $E^2/(GM^3) = g$, between the total energy E and the total mass M, we find:

$$E \equiv (gG)^{\frac{1}{2}} M^{\frac{3}{2}}.$$

Since $(gG)^{\frac{1}{2}}$ is a constant, it can be seen that the dispersion of energy grows with a factor of $M^{\frac{3}{2}}$, while the mass M grows with a factor of M^1. It sets an upper limit to the size of a galaxy under normal conditions. This can also be seen from the tight relation between the size of a black hole and the brightness of a galaxy, that entails that the greater the black hole becomes, the more material the galaxy spews out. If this connection did not set an upper limit to the size of the galaxies, they would, due to gravity, grow out of all proportion.

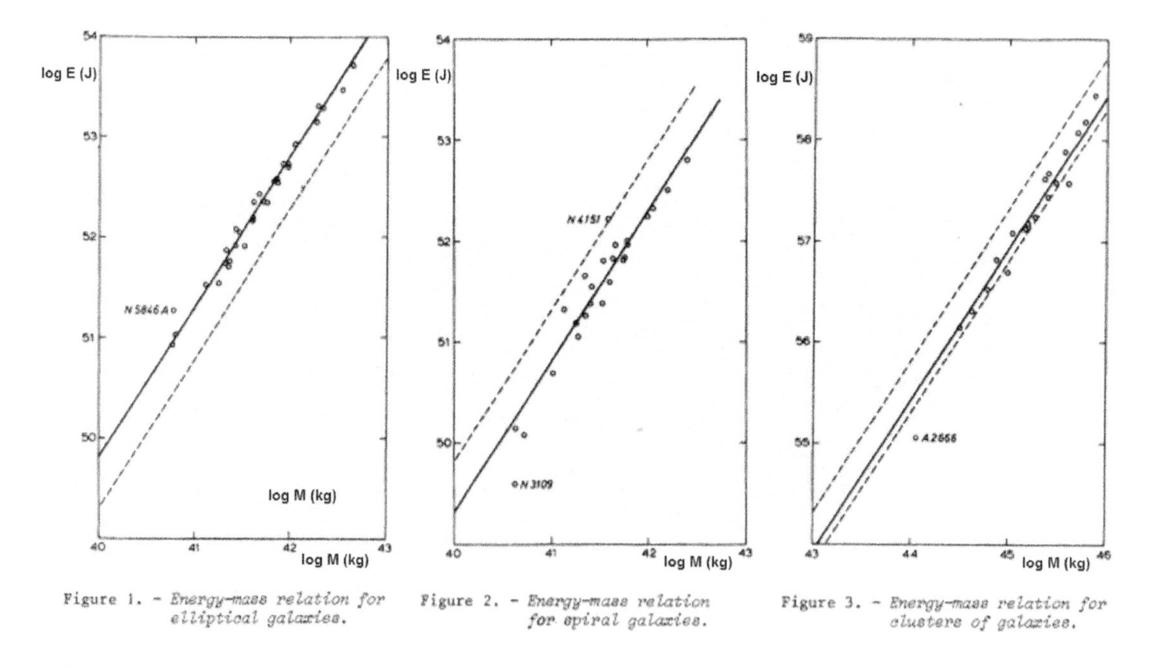

Figure 3.7: Energy-mass relations for galaxies. [143]

However, if a galaxy for some reason receives an excess of matter, such as when two galaxies merge together, the size can become so big that the black hole at the center of the galaxy literally tears itself apart. [145] This has been observed in a far-off galaxy, 12.4 billion light-years from the Earth. The observation of the fate of the galaxy called W2246-0526 is based on data from NASA's Wide-field Infrared Survey Explorer (WISE). At the time of the explosion it had the highest power output of any galaxy in the Universe. Moreover, it would appear to shine as the brightest galaxy, if all the galaxies were at the same distance from us. [146]

3.5.3 The life cycle of Matter and Radiation

As it can been seen there is an energy cycle, where SMBHs emit huge amounts of matter and radiation, which later end up in the interstellar medium and blend smoothly with the intergalactic space. Most of the interstellar medium consists of gas, dust, cosmic rays and electromagnetic radiation, [147] where the most common interstellar gas is ionized hydrogen and helium from the black holes. The densest regions of the interstellar medium are the molecular clouds where the stars are formed.

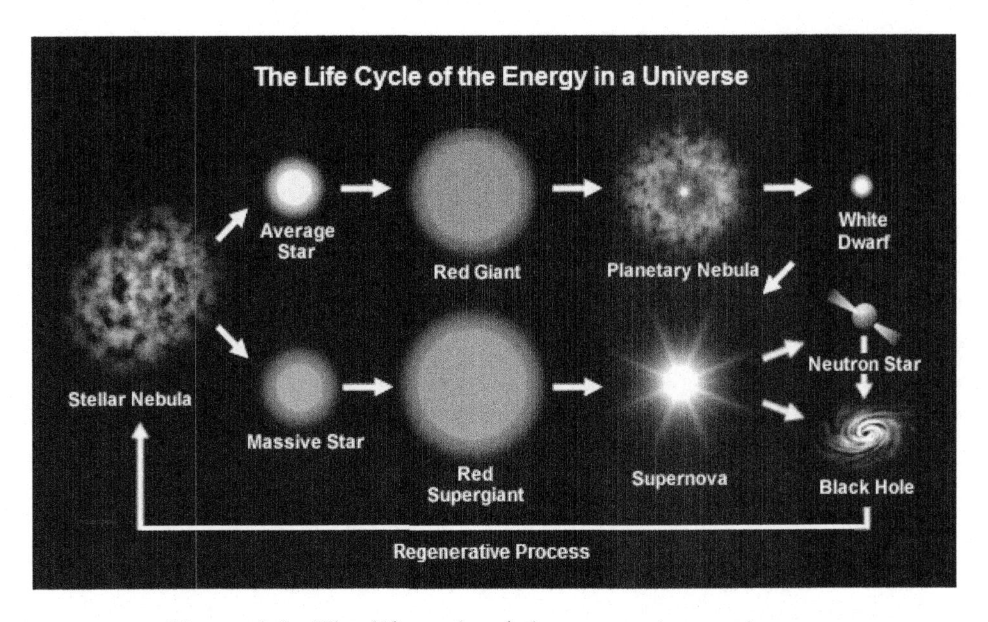

Figure 3.8: The life cycle of the energy in a universe.

During the period from the formation of a molecular cloud to the creation of a black hole, the stars help replenish the interstellar medium with matter and energy by means of planetary nebulae, stellar winds, and supernovae. Planetary nebulae is ejected from old red giant stars, [146] and consists of expanding, glowing shells of ionized gas, which emit electromagnetic radiation of various frequencies, while the stellar core remnants may ultimately end up as a white dwarfs. The stellar winds are streams of charged particles that mostly consist of electrons, protons and alpha particles with thermal energies between 1.5 and 10 keV, which stem from the coronae of Sun-like stars.

A supernova expels much of its material away from the star, at velocities up to 30,000 km/s. [76] This drives an expanding and fast-moving shock wave into the surrounding interstellar medium, which sweeps up a growing shell of gas and dust, which is later observed as a supernova remnant. [148] The supernovae create, fuse and eject the bulk of the chemical elements produced by nucleosynthesis, and thereby play a significant role in enriching the interstellar medium with heavier elements. [149], [150]

The intergalactic space is the physical space stretching between the galaxies, which contains a rarefied plasma that is organized in a galactic filamentary structure. [151], [152] The material is called the intergalactic medium and has a density equal to 5-200 times the average density of the Universe. [153] It primarily consists of ionized hydrogen. Since gas is continually supplied to the intergalactic medium, and since the cosmic microwaves transfer energy to the medium, (ch. 2.5.1) the microwaves become red-shifted, while the medium is heated to a temperature between 10^5 K and 10^7 K, [154] which is high enough to ionize the hydrogen, which is why the intergalactic medium is ionized.

At these temperatures, it is called the warm–hot intergalactic medium (WHIM), and observations indicate that up to half of the atomic matter in our Universe might exist in this warm–hot, rarefied state. [153], [46], [155] When gas gravitates from the filamentary structures of the WHIM, towards the galaxy clusters at the intersections of the cosmic filaments, it can heat up reaching temperatures of 10^8 K and above, in the so-called intra-cluster medium. [156] However, some of the matter and radiation may also try to leave the intergalactic space, but since it cannot leave the closed Universe, it is forced to turn around.

3.5.4 The Largest Galaxies produce the Largest Eruptions

The size of a galaxy gives an indication of the size of the black hole at the center, and the extent to which it is able to create a regenerative process. Since larger galaxies attract more of the intergalactic matter and radiation, the larger galaxies grow faster than the smaller ones, and they may often grow by merger with other galaxies. This is for instance the most likely fate of the Milky Way, which is supposed to merge with the nearby galaxy, M31, that lies in the constellation of Andromeda. The end result will most probably be a much larger elliptical galaxy, depending on the collision angle.

On the other hand - the larger the galaxies become, the more frequent and powerful are the eruptions, so the more matter and energy they disperse. In the centers of almost all known massive galaxies there is a SMBH. [144] The largest types of black holes are in the range of 10^5 to 10^9 $M_\odot$, [157] and in the case of the Milky Way, the SMBH is Sagittarius A*, which is a light and very compact astronomical radio source in the middle of our galaxy. [158]

Black holes with a mass greater than the Chandrasekhar limit of 1.39 solar masses [75] and without a significant thermal pressure will collapse due to the external pressure to generate a neutron star at the center. This implies that all black holes at the center of the galaxies, solely due to their size, contain a neutron star. [143] The neutron stars at the center of the galaxies have a long lifespan, during which the neutron stars often start as fast rotating stars with a very potent magnetic field. When the neutron star collapses, its rotation rate increases as a result of conservation of the angular momentum, so newly formed neutron stars rotate up to several hundred times per second; and the magnetic field strength on the surface of the neutron stars are estimated to have a range of at least 10^4 to 10^{11} tesla, [105] and are observed to have a surface temperature around 6×10^5 K. [99]

A black hole with a neutron star at the center will as a consequence of accretion of matter generate an outer pressure on the neutron star which, together with quantum fluctuations, will be able to overcome the Tolman–Oppenheimer–Volkoff limit and break apart the neutrons into their constituent quarks and gluons. This process will, due to the huge difference between the volume and mass of the neutrons, and the volume and mass of the resulting quarks, generate a gravitational collapse combined with an explosive release of large amounts of energy, such as electromagnetic radiation and a plethora of different particles. Depending on the magnetic field and the rotational speed of the neutron star, such a regenerative process may take many different forms. [159]

The accretion, inner shells, size, and rotation speed of the neutron star and the black hole, together with the direction and strength of the magnetic field in relation to the axis of rotation, are the main factors that determine the type and periodicity of the eruptions. The radiation from pulsars is thought to be emitted from a region near the magnetic pole, so if the magnetic pole does not coincide with the rotational axis of the neutron star, the emission beam will sweep the sky. From the treatment of neutron stars we know that the bursts contain a sea of quark-gluon plasma, which rapidly transforms into nucleons and electrons, in addition to the visible X- and gamma-rays. Finally, the nucleons most probably end up as ionized hydrogen and helium in the stellar nebulae, where they participate in the formation of new stars. [160]

Outside the centers of the galaxies stellar black holes exist with masses ranging from about 5 to between 10-100 solar masses [161] that can generate hypernova explosions, gamma rays, or X-ray bursts. [159] Such bursts can occur several times a day and last for about ten seconds each. But the fusion of hydrogen into helium, carbon, oxygen, silicon and iron may also release enormous amounts of energy. [162]

Slow-rotating non-accreting neutron stars are almost undetectable, and when they reach a mass of 3.0 $M_\odot$. [17] they become black holes. So - since all the stars at the end of their life cycle end up as black holes before they participate in a regenerative process - black holes constitute along with black dwarfs and barren neutron stars the majority of dark matter in the galaxies. In spiral galaxies, such as our own, the mass distribution in the plane perpendicular to the axis of rotation follows the centrifugal principle. In centrifugation, the force exerted by rotational motion separates the components in relation to their mass, so the lighter particles, relative to the mean value of the mass of the particles, approach the center of the galaxy, where they supply the fuel to the luminous bulge mass, while the heavier black holes are flung towards the periphery, where they constitute most of the dark matter, called the halo.

3.5.5 Regenerative Processes in Neutron Stars and Black Holes

A black hole with a neutron star at its center is often named an active galactic nucleus (AGN), and a galaxy that hosts an AGN is called an active galaxy. The radiation from an AGN is the result of accretion of matter by the SMBH. These AGNs are the most luminous persistent sources of radiation in the Universe, and the area around an AGN is a compact region at the center of a galaxy, which has a much higher than normal luminosity throughout the entire electromagnetic spectrum, with characteristics indicating that the extreme luminosity is not produced by stars. Such emissions have been observed in the radio, microwave, infrared, optical, ultra-violet, X-ray and gamma-ray wavebands.

There are plenty of old and cold neutron stars and black holes in the galaxies; however, it is the neutron stars and black holes at the center of the galaxies that receive most of the fuel. The most common eruptions of neutron stars are "soft gamma repeaters", which emit large bursts of X-rays and gamma-rays at irregular intervals, and depending on the rotational speed and the strength of the magnetic field of the neutron star the eruptions can vary from steady X-ray beams to gigantic explosions. 'X-ray pulsars' consist of neutron stars in orbit with normal stellar companions, and 'millisecond pulsars' are pulsars with a rotational period in the range of about 1-10 milliseconds, which have been detected in the radio, X-ray or gamma-ray area of the electromagnetic spectrum. Besides this, a wide range of different regenerative processes occur such as type-I X-ray bursts, [163] type-II X-ray bursts, [164] superbursts, [165] and gamma-ray bursts, [125] which all are connected to neutron stars and among the most powerful explosions.

The characteristics of an AGN such as the presence or absence of jets, depend on several properties such as the mass of the central black hole, the rate of accretion onto the black hole, its rotational speed and the strength of the magnetic field. If the black hole has a high rotational speed and a powerful magnetic field, the emissions will leave the black hole in an orderly way, which can be two diametrically opposing beams with angles of only a few degrees. The cosmic rays from AGN and supernova explosions primarily consist of high-energy protons and atomic nuclei, [166] and the highest-energy fermionic cosmic rays observed to date had an energy of about 3×10^{14} MeV, while the highest-energy gamma rays are photons with energies of up to 10^8 MeV. [167], [168]

AGNs can be Seyfert galaxies or quasars, which are the two largest groups of active galaxies, where quasars are classified as the most powerful. AGN galaxies such as radio galaxies are known as Seyfert galaxies in honor of Seyfert's pioneering work identifying the visible light sources associated with the radio emission. In photographic images, some of these objects were nearly point-like or quasi-stellar in appearance, and were classified as quasi-stellar radio sources (later abbreviated as "quasars"). Early X-ray astronomy observations demonstrated that Seyfert galaxies and quasars are powerful sources of X-ray emission, which originates from the inner regions of massive black holes at the center of the galaxies (10^6 to 10^{10} times the solar mass). [169]

A quasar is an active galactic nucleus that consists of a black hole with a neutron star at the center, which is surrounded by an orbiting accretion disk. The quasars are placed at the center of the galaxies, which delivers the flow of material to the accretion disks. Quasars are found over a very broad range of distances, from redshifts of roughly 0.1 to 7, which is just from the neighborhood to as far as we can currently see. Quasars inhabit the center of almost every galaxy, and are among the most luminous, powerful, and energetic objects known in the Universe, emitting up to a thousand times the energy output of the Milky Way. This radiation is emitted across the whole of the electromagnetic spectrum, and is almost uniform in strength from X-rays to the far-infrared with a peak in the ultraviolet-optical bands. Some quasars are also strong sources of radio emissions and gamma-rays.

Based on the burst duration and release of energy, gamma-ray bursts can be roughly divided into two classes, the short and the long gamma-ray bursts. [170] The short gamma-ray bursts tend to have harder spectra than the long gamma-ray bursts. The total energy released in a short gamma-ray burst during the first hundred seconds is $\sim 10^{43}$ joule, while the long gamma-ray bursts are a few hundreds to a few thousand times more energetic, [171] which fits well with the energy production of a neutron star.

Gamma-ray bursts or super-bursts are much rarer than ordinary, helium-powered bursts, and release a thousand times more energy. The super-bursts are so long and bright that they act as spotlights beamed from the surface of the neutron stars. However, gamma-ray bursts are the brightest electromagnetic events known to occur in the Universe. [172] The gamma-ray bursts can last from ten milliseconds to several hours. [173] After an initial flash of gamma rays a longer-lived afterglow is usually emitted at longer wavelengths. [174] Gamma-ray bursts are released from a rapidly rotating, high-mass neutron star or a black hole. A subclass of gamma-ray bursts called "short" bursts is so rare that there are only a few per galaxy per million years. A typical burst releases just as much energy during a few seconds as the Sun in its entire 10-billion-year lifetime. These short gamma-ray bursts have all been observed in distant galaxies. [175]

The galaxies generally live in harmony with each other, where every galaxy has its own life, which is only disturbed by collisions with other galaxies. The energy circuits of the galaxies mean that the galaxies, through a series of nuclear reactions, create ever-heavier elements, which during a life cycle, at last reach the center of the same or other galaxies where they are decomposed and ejected as nucleons, electrons and photons. Looking outside our galaxy we see structures on all scales, from dwarf galaxies to galaxy clusters, and sheet-like structures of galaxies separated by enormous voids.

3.6 The Maximum Size of our Universe

According to the Euclidean Cosmos Theory the Cosmos contains a finite number of universes that all are almost infinitely old. So each of the universes have had plenty of time to settle in relation to the forces that act on their constituent matter. The most important forces in relation to the distribution of matter are gravity and the more or less explosive energy production. So in each of the closed universes there arises, just as in the galaxies and stars, an equilibrium between the inward pull from the center of mass and the outward expansion from the generation of energy. So we can presume that the distribution of mass in a universe is equal to the distribution of the mass in the Sun. When the Cosmos has settled down, the space between the universes can only obtain matter and energy through a merger of universes, so the space between the universes must be almost depleted of matter, which is why the maximum size of a Universe is determined by whether it is able to hold on to its energy at a given mass distribution, which it is, when it is closed.

Figure 3.9: The mass distribution in our spherical Universe.

Due to gravity and the expanding explosive forces, all the universes are expected to be spherical, with a gradually falling mass density from the center to the circumference with radius R. From the density of ordinary matter in our Universe, and the condition that our Universe is closed, it is then possible to make an estimation of the maximum size of our Universe. Since the radius, R, of the Universe, must be less than or equal to the Schwarzschild radius (see eq. 2.48), we have:

$$R \leq r_{Schwarzschild} = \frac{2GM}{c_0^2} ,$$

where G is the gravitational constant, M is the total mass of our Universe and c_0 is the speed of light.

Consider the gravitational force that a spherical mass distribution exerts on a mass dm located at the distance r from the center of a universe. Since the contribution outside the sphere with radius R, where R is equal to the Schwarzschild radius, does not exist, it is only the interior mass $M(r)$ that contributes to the gravitational attraction:

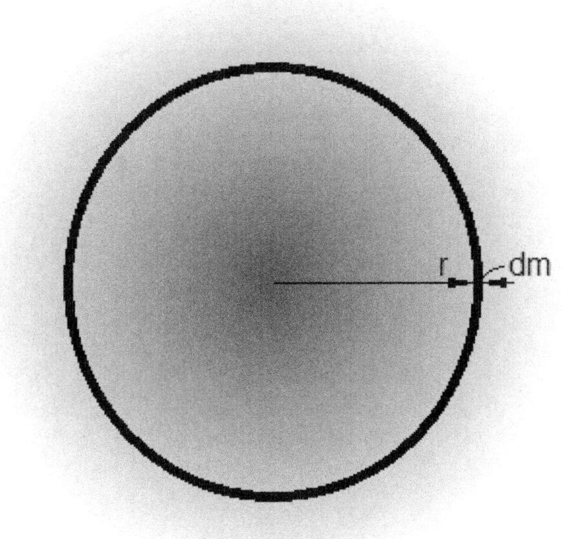

Figure 3.10: The mass distribution in our spherical Universe.

As the universes are spherically symmetrical, the total mass of a universe M can be expressed by the sum of the volumes of concentrically spherical shells, $dV = 4\pi r^2 \cdot dr$, with radius r and the thickness dr, times their density $\rho(r)$, which is dependent on the distance from the center, so the total mass of a universe equals:

$$M = \int_0^R \rho(r) dV = \int_0^R 4\pi r^2 \cdot \rho(r) \cdot dr .$$

Since the density is a function of the radius $\rho(r)$, and a universe like a spherical star gets its mass distribution from the influence of gravity and the nuclear activity, there is much to suggest, that a universe has a mass distribution like the Sun. [176] So for the moment the best estimate for the density ρ is:

$$\rho = \rho_0 \cdot e^{-9r}, \text{ for: } 0 \leq r \leq 1,$$

where the graph for the density is shown below:

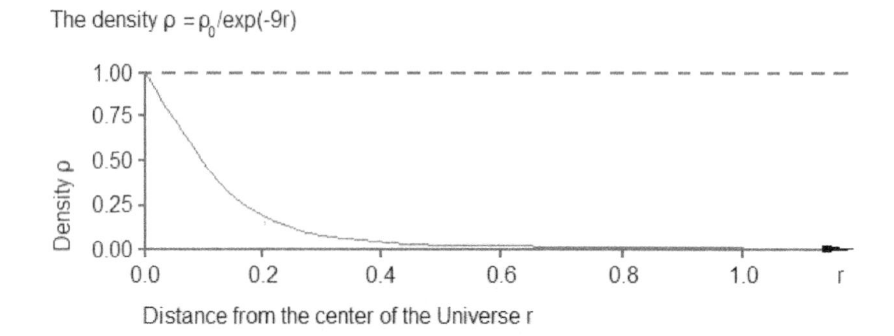

Figure 3.11: The density as a function of the radius of the Universe.

When we insert the density function in the expression for the total mass of a universe, we get:

$$M = \int_0^R \rho(r/R)dV = \int_0^R 4\pi r^2 \rho_0 \cdot e^{-9r/R} \cdot dr = 4\pi \cdot \rho_0 \int_0^R r^2 \cdot e^{-9r/R} dr$$

so:

$$M = 4\pi \cdot \rho_0 \Big[-\frac{R}{729} e^{-9r/R} (81r^2 + 18Rr + 2R^2) \Big]_0^R$$

$$= 4\pi \cdot \rho_0 \Big(-\frac{R}{729} e^{-9} (81R^2 + 18R^2 + 2R^2) + \frac{R}{729} 2R^2 \Big)$$

$$= 4\pi \cdot \rho_0 \frac{R^3}{729} (2 - 101e^{-9}) .$$

As the universes are closed, the maximum radius R must be equal to:

$$R = \frac{2GM}{c_0^2} = \frac{8G \cdot \pi \cdot \rho_0 R^3 (2 - 101e^{-9})}{729 c_0^2}, \text{ from which}$$

$$R = \sqrt{\frac{729c_0^2}{8G \cdot \pi \cdot \rho_0(2 - 101e^{-9})}}\,.$$

Recent observations of our Universe have shown that the density of ordinary matter such as atoms accounts for about 4.6% of the critical density $\rho_c = 9.9 \times 10^{-27}$ kg/m^3. [177] So the density of ordinary matter is equal to $\rho_0 = 4.6 \times 10^{-28}$ kg/m^3.

Since the Universe is closed, the gravitational field from the Universe cannot hold on to radiation outside a sphere of radius R_o, where $G = 6.67 \times 10^{-11}$ m^3 kg^{-1} s^{-2}, $c_0 = 3.00 \times 10^8$ m/s and the density of ordinary matter is $\rho_o = 4.6 \times 10^{-28}$ kg/m^3. So, if the density of the Universe is equal to the density of ordinary matter, the estimated radius of the Universe equals:

$$R_o = \sqrt{\frac{729c_0^2}{8G \cdot \pi \cdot \rho_0(2 - 101e^{-9})}} =$$

$$\sqrt{\frac{729 \cdot (3.00 \times 10^8)^2}{8 \cdot 6.67 \times 10^{-11} \cdot 3.14 \cdot 4.6 \times 10^{-28}(2 - 101e^{-9})}} = 6.5 \times 10^{27} \text{ m}\,.$$

However, it is much more realistic that the total amount of matter is equal to ordinary matter plus dark matter, where the density of dark matter accounts for about 24% of the critical density $\rho_c = 9.9 \times 10^{-27}$ kg/m^3. [177] So the density of ordinary matter plus dark matter is $\rho_{o+d} = 28.6 \times 10^{-28}$ kg/m^3. That is to say, if the density of the Universe is equal to the density of ordinary matter plus dark matter, the estimated radius R_{o+d} of the Universe equals:

$$R_{o+d} = \sqrt{\frac{729c_0^2}{8G \cdot \pi \cdot \rho_{o+d}(2 - 101e^{-9})}} =$$

$$\sqrt{\frac{729 \cdot (3.00 \times 10^8)^2}{8 \cdot 6.67 \times 10^{-11} \cdot 3.14 \cdot 28.6 \times 10^{-28}(2 - 101e^{-9})}} = 2.6 \times 10^{27} \text{ m}\,.$$

By comparison, the radius of our observable Universe is equal to 4.4×10^{26} m, [57] so Dark Energy (if any, since our Universe does not expand) must be matter outside the visible Universe, while the Dark Matter is the old invisible matter inside the visible Universe such as black dwarfs, barren neutron stars and black holes.

We can now make an estimation of the mass of ordinary matter M_o in our Universe:

$$M_o = 4\pi \cdot \rho_0 \frac{R^3}{729}(2 - 101e^{-9}) = 4 \cdot 3.14 \cdot 4.6 \times 10^{-28} \cdot \frac{(6.5 \times 10^{27})^3}{729}(2 - 101e^{-9}),$$

which is equal to:

$$M_o = 4.3 \times 10^{54} \text{ kg}.$$

In the same way we can find the estimated mass of our Universe including dark matter:

$$M_{o+d} = 2.7 \times 10^{55} \text{ kg}.$$

In calculating the size of our universe, we have tacitly assumed that our observable Universe is located at the center of our Universe. However, this is not the case, as it has been observed that the observable Universe is lopsided (Fig. 4.8). So the density of our Universe must be expected to be somewhat larger, thereby making our Universe larger than the calculated size. From the density profile of our observable Universe, we may even be able to determine where we are located in our Universe.

3.6.1 The Limit of the Visible Universe

Even the strongest light sources fade away when the distance increases, which is due to parameters such as redshift, reflection, refraction, diffraction and absorption, besides the sheer distance of the sources, as the intensity of light diminishes with the square of the distance. Consequently, the limit of the visible Universe is determined by how far we can see with the most powerful telescopes, which again depends on the brightness of the objects, the spread of light as it travels through space, and the risk of distortion and dispersion.

To evaluate the amount of light from a remote light source, the flux F from the distant source can be determined, where the flux equals the total amount of energy from the source that crosses a unit area per unit time. The flux is measured in joules per second per square meter ($\text{Jm}^{-2}\text{s}^{-1}$) or watts per square meter (W/m^2).

To determine the flux from an astronomical object, photometry is used. [178] Photometry is the branch of physics concerned with the measurement of the total flux or intensity of an astronomical object. The methods used to perform photometry depend on the wavelength of the object's electromagnetic radiation, and can for instance be conducted by gathering light in a telescope and measuring the energy with a photosensitive instrument.

Figure 3.12: The light we receive from the galaxies diminishes with the square of the distance.

To determine the amount of energy from a distant source, we may first measure the luminosity L. The luminosity is the total amount of energy emitted by an astronomical object per unit time and is measured in joules per second, or watts, W. [179] Since the luminosity is the total amount of energy per unit time, the flux must be equal to the luminosity per unit area A:

$$F = L/A,$$

Consequently, if an astronomical object with luminosity L radiates equally in all directions, the observed flux F at the distance r from the source, will be equal to the luminosity per unit area at the distance r:

$$F = \frac{L}{4\pi r^2}, \tag{3.6}$$

where F is the measured flux at the observer in $\mathrm{W/m^2}$, L is the luminosity of the source in watts, and r is the distance from the source to the observer in meters.

Among the most distant observed astronomical objects is the galaxy, GN-z11, which is 13.39×10^9 light-years (ly) away, [180] the galaxy EGSY8p7, which is 13.23×10^9 ly away, [181] a gamma-ray burst 13.18×10^9 ly away, [182] and a quasar 13.05×10^9 ly away. [183] As it can be seen from the examples, all the faintest astronomical sources have faded away with the distance, so only the brightest galaxies and quasars are visible.

Accordingly, the radius of the visible Universe is around 13.39×10^9 ly. [180]

Let us as an example take a closer look at the galaxy GN-z11 that emits ultraviolet waves in the emission spectrum of atomic hydrogen, namely the Lyman-alpha emission line of hydrogen with the wavelength of $\lambda_s = 121.6$ nm, [180] where the subscript 's' refers to the source. Due to the plasma redshift, the ultraviolet light from GN-z11 has lost some of its energy to the CMB, so the observed wavelength is equal to $\lambda_o = 1470$ nm, [180] in the near-infrared, where the subscript 'o' refers to the observer.

It was the Hubble Space Telescope that found GN-z11. [180] The galaxy is unexpectedly luminous for a galaxy at that distance, as its ultraviolet luminosity L_s is 3 times larger than the characteristic galaxy luminosity L_*, which is roughly comparable in luminosity to the Milky Way, [184] which in turn is equal to $L_* = 1.4 \times 10^{10} \cdot L_\odot$, [180] where $L_\odot = 3.8 \times 10^{26}$ W is the luminosity of the Sun. [185] So, the ultraviolet luminosity of the galaxy GN-z11 equals:

$$L_s = 3 \cdot L_* = 3 \cdot 1.4 \times 10^{10} \cdot L_\odot = 3 \cdot 1.4 \times 10^{10} \cdot 3.8 \times 10^{26} \text{ W} = 1.6 \times 10^{37} \text{ W}.$$

As the ultraviolet light from the source GN-z11 with the wavelength $\lambda_s = 121.6$ nm loses some of its energy to the CMB, the observed wavelength is equal to $\lambda_o = 1470$ nm, which is in the near-infrared spectrum.

From Planck's radiation law we know that the energy E per photon with the frequency ν is equal to $h\nu$, where h is Planck's constant. So, the total number of emitted ultraviolet photons per second, N_s/s, with the frequency, ν_s, times Planck's constant, h, must be equal to the luminosity of the source GN-z11, that is $L_s = N_s h\nu_s/s$. Likewise, the total number of observable near-infrared photons per second, N_o/s, with the frequency, ν_o, times Planck's constant, h, must be equal to the luminosity of the observable photons, so $L_o = N_o h\nu_o/s$, where $N_s = N_o$, if none of the photons are lost. It is now possible to find an expression for the luminosity of the observable near-infrared light. Since the luminosity of the source GN-z11 is equal to:

$$L_s = N_s h\nu_s/s \Rightarrow N_s h/s = L_s/\nu_s$$

and since the luminosity of the observable light is equal to: $L_o = N_o h\nu_o/s$ and the frequency $\nu = c_0/\lambda$, we get:

$$L_o = N_o h\nu_o/s = \nu_o N_s h/s = L_s\nu_o/\nu_s = L_s(c_0/\lambda_o)/(c_0/\lambda_s) = L_s\lambda_s/\lambda_o$$

So, the luminosity of the observable near-infrared light is equal to:

$$L_o = L_s\lambda_s/\lambda_o = 1.6 \times 10^{37} \cdot 121.6/1470 \text{ W} = 1.3 \times 10^{36} \text{ W}$$

However, it will only be a small part of the photons from the galaxy GN-z11 that reach the Hubble Space Telescope, as it is placed at the observable limit of our Universe, 13.39×10^9 light-years away. [180] Since a light-year is around 9.46×10^{15} m, the distance to the galaxy is about 1.3×10^{26} m, so the flux density that reaches the observer on the Earth equals:

$$F = \frac{L_o}{4\pi r^2} = \frac{1.3 \times 10^{36}}{4\pi(13.39 \times 10^9 \cdot 9.46 \times 10^{15})^2} = 6.4 \times 10^{-18} \text{ W/m}^2.$$

As the received light from the galaxy GN-z11 is near-infrared photons with a wavelength equal to $\lambda_o = 1470$ nm, with an energy equal to:

$$E_{\lambda_o} = h\nu_o = hc_0/\lambda_o = 6.63 \times 10^{-34} \cdot 3.0 \times 10^8/1470 \times 10^{-9} \text{ J} = 1.35 \times 10^{-19} \text{ J},$$

the Hubble Space Telescope will only receive:

$$N_o = F/E_{\lambda_o} = \frac{6.4 \times 10^{-18}}{1.35 \times 10^{-19}} = 47 \text{ photons/m}^2 \text{ per second},$$

from the galaxy GN-z11. Here we have used Planck's constant $h = 6.63 \times 10^{-34}$ kg m^2/s and the speed of light $c_0 = 3.0 \times 10^8$ m/s.

The image of GN-z11 has pushed the Hubble Space Telescope to its limits and increased the cosmic distance record by measuring the distance to the most remote galaxy ever seen in the Universe. GN-z11 was found using the Wide Field Camera 3 (WFC3), which is the fourth-generation UVIS/IR imager aboard the Hubble Space Telescope, where UVIS is an abbreviation of Ultraviolet Imaging Spectrograph, and IR stands for Infrared Radiation. It is the most technologically advanced instrument to take images in the visible spectrum.

The galaxy GN-z11 only became visible for Hubble because of its high activity, which causes the galaxy to be unusually bright compared to its distance. This is the first time that the distance of an object so far away has been measured from its spectrum, which makes the measurement extremely reliable. [186] The Hubble Space Telescope's takes images with a 2.4-meter mirror, and has four main instruments in the near ultraviolet, visible, and near infrared spectra. Hubble's orbit allows it to take extremely high-resolution images, with substantially lower background light than ground-based telescopes.

When we reach the limits of the photometric units, it will only be the centers of the large galaxies that will be luminous enough to be observed by the photometric units, while the outer regions of the galaxies will act as dark matter. That is why we find most quasars at the observable limit of our Universe. Another parameter that sets an upper limit for our perception of distant objects, and thereby an upper limit for the observable Universe, is the plasma redshift, since it transforms the light so it becomes dimmer and darker.

3.7 The Large-Scale Structures of the Universes

Since each of the universes, except for the collisions, have existed for an infinite length of time, the universes have had plenty of time to generate the large-scale structures through the accretion of matter, and the accumulation of matter has created the basic structure of the galaxies with their large black holes at the center.

Figure 3.13: The network structure in our part of the Universe. [187]

When, due to accretion, the black holes get a mass greater than between 8 and 29 $M_{\odot}$, they may generate a neutron star at the center, if it was not there in the first place. Through accretion of matter the center of the black holes will then from time to time get a density which surpasses the Tolman–Oppenheimer–Volkoff limit of neutron degenerate matter of 3.2 $M_{\odot}$, so a black hole periodically becomes an AGN, which creates different types of electromagnetic radiation, electrons and nucleons. Some of the radiation and elements are scattered all over the universe, where the nucleons end up as a part of the intergalactic medium, which contains around half of the atomic matter in a universe, [153], [46], [155] while the rest may end up as nebulae from which the stars are born that illuminate the universe and provide the galaxy with a range of different elements.

When the stars have ended their life cycle they will most likely end up as dark matter since it only requires 0.08 $M_\odot$ to create a black dwarf, [188] 1.1 $M_\odot$ to create a barren neutron star, [78] and around 3 $M_\odot$ to create a black hole, [17] so the universes will be teeming with such dark objects. But since neither a black dwarf, a barren neutron star or a black hole emits any radiation, as long as they do not receive any fuel in the form of matter, they are only noticeable through the gravitational pull they exert on the galaxy, which is striking when considering the velocity distribution of the galaxies.

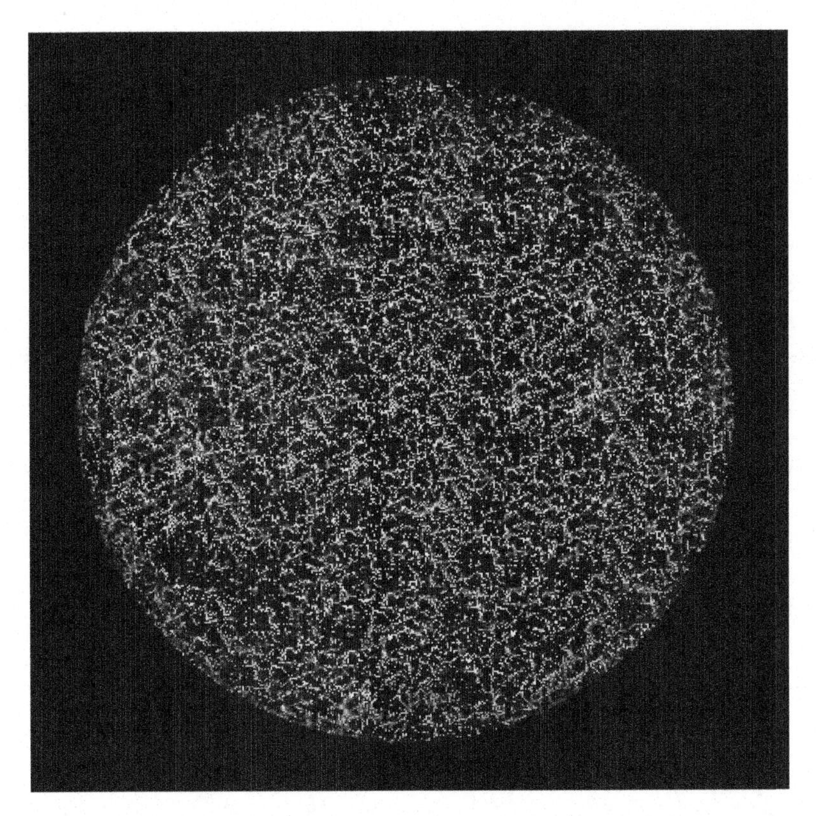

Figure 3.14: The Large-scale Structure of the Universe.

Due to gravity, the galaxies assemble into ever larger structures, which make up the frameworks of the universes with their filaments and great walls. In the large structures formed due to the law of least effort, the galaxies distribute themselves as the structure of a honeycomb, where most of the galaxies, due to gravity, gather at the intersections. During their active periods, all these galaxies will spread their energy as photons, electrons and nucleons, which end up as cosmic rays and the intergalactic medium.

Since each of the universes is almost static, there has been plenty of time to create the great walls. However, even the great walls will gravitate against the most powerful center of mass. At places where the density becomes greater, the galaxies will collide and create still heavier galaxies. But the larger the galaxies get, the greater the black holes at their centers become, which then by means of regenerative processes will spread the matter and energy toward the farthest corners of the universe. Since the universes are closed they will in this way be able to maintain a stable distribution of matter and energy. However, if a closed universe of some reason loses the grip on some of its energy, such as when, for instance, a barren object, a black hole or another universe passes close by, the energy may be attracted to the object and end up in, for example, the other universe.

Based on observations of redshift, gigantic structures have been found, where galaxies and clusters of galaxies are distributed in network-like structures extending over several hundred million light-years. For instance, there has been found a sheet of galaxies called the "Great Wall", which is more than 500 million light-years long and 200 million light-years wide, but only 15 million light-years thick; and one of the largest gaps ever observed has a diameter of about 230 million light-years. [189]

The huge Sloan Great Wall spans over one billion light-years, and the Coma cluster is one of the largest observed structures in the Universe, containing over 10,000 galaxies and extending more than 1.37 billion light-years in length. [190] Hundreds of millions of galaxies have clumped together, forming super clusters and series of massive walls of galaxies separated by vast voids of empty space. Some of these elongated super clusters have formed a series of walls, one after another, spaced from 500 million to 800 million light-years apart, such that in one direction alone, 13 great walls have formed with the inner and outer walls separated by about seven billion light-years. Recently, cosmologists have estimated that some of these galactic walls may have taken from 80 billion to 150 billion years to form. [191]

3.7.1 The Large-scale Structure is Mirrored in the CMB

Among the most active galaxies in our Universe are the quasars, and since they reflect the grade of activity, they can be used as lighthouses for the distribution of mass concentrations in our Universe. From the activity (Fig. 3.15) it is seen that the area of the sky which is not hidden by our own galaxy shows a very uneven distribution of quasars. These regenerative processes are also mirrored in the CMB.

The CMB that stems from the regenerative processes provides some of the best knowledge we have about the structure and content of our Universe (Fig. 3.16). Smaller details in the fluctuations tell us about the relative amounts of matter, such as black holes, galaxies and other celestial bodies. And some of the largest fluctuations - covering one-sixteenth, one-eighth, and even one-fourth of the sky - mirror the greatest structures.

From the CMB it can likewise be seen that our Universe has certain dark and bright spots on a very large-scale. These large-scale fluctuations in the power spectrum reflect the vast voids of empty space separated by the great massive walls of galaxies, where the regenerative processes take place, and the temperature variations are just as large as one

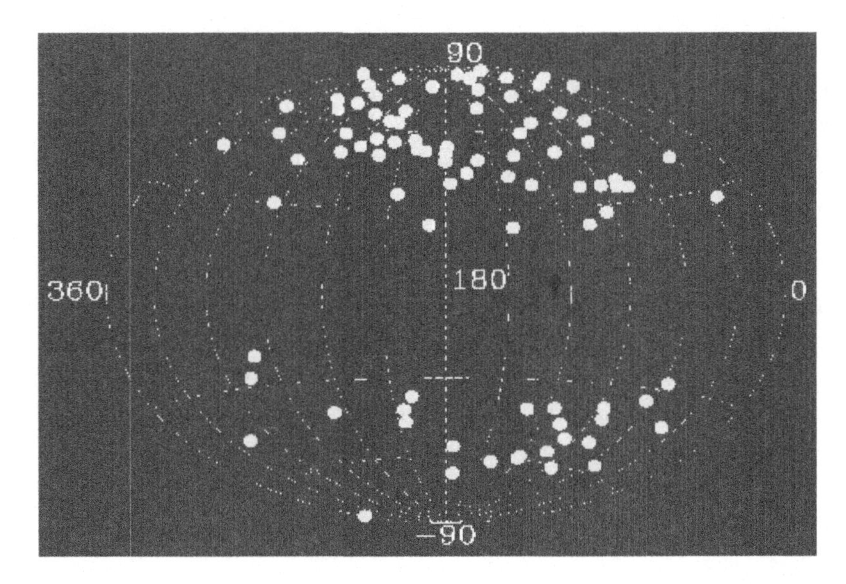

Figure 3.15: Groups of quasars arise where the mass density is highest.

would expect when there is a continuous series of regenerative processes. The physicist Erich Regener, [192] used the total measured energy of cosmic rays to estimate the intergalactic temperature of the cosmological background radiation to 2.8 K. Compared with the latest estimates of the temperature of the CMB at 2.72 K, [193] it can be seen, that the temperature derive from the regenerative processes.

Figure 3.16: All-sky map of the cosmic microwave background.

Although the temperature of the cosmological background radiation is almost constant, which is due to the deflection of photons as our Universe is closed, the structure of the Universe is very uneven. [187] Based on angular positions and redshift surveys of sections of the sky, which are used to calculate the distance of astronomical objects from the Earth, gigantic structures have been found everywhere in the visible Universe. [189]

Finally, it can be seen from the theory that our Universe is flat, which means that it is Euclidean, and that it has the property that it will neither expand nor collapse. This property is fulfilled because of the balance between the activity of the regenerative processes that are constantly spreading matter and energy and the gravitational attraction that is constantly trying to gather it. The fact that our universe has a flat geometry is also confirmed by the *Omega* density parameter, which is very close to one, ($\Omega = 1.02 \pm 0.02$), [6] where the density parameter is the ratio of the total density ε and the critical density ε_C, which is the density at which our Universe is completely flat. [6]

Chapter 4

Discussion

4.1 Evaluation of the Quantum Ether Theory

All the relativistic connections which are deduced in connection with the Quantum Ether Theory are based on the existence of the ZPF, which is a combination of the lowest energy states of all fields. This is in line with Hendrik Antoon Lorentz's deduction of the Lorentz contraction from the existence of an ether, [2] where the relativistic relations likewise arose from a motion relative to the propagation velocity of the electromagnetic field.

4.1.1 The Space is Euclidean

In connection with the deduction of the Euclidean Cosmos Theory, it is crucial that the space is Euclidean. When we consider the expressions for the length contraction $x' = x\sqrt{1 - v^2/c_0^2}$ and the "time dilation" $t' = t\sqrt{1 - v^2/c_0^2}$ - which occurs in Einstein's theory of relativity (ref. [7] p. 48) - it can be seen that the time is shrinking by exactly the same factor as the length, so the time passes just as fast in a moving system as in a stationary system, which means that the time is absolute and universal. When we look at space and time as a combined space-time, the time axis is thus just as linear as the three coordinate axes, and since the curvature of space-time only depends on the time axis, the space does not curve. It means that the space is Euclidean, so the gravitational field cannot be explained by the curvature of space-time.

Just as the other three fundamental forces are generated by virtual particles, it is obvious that the gravitational force is also generated by virtual particles, which can be seen from the fact that the gravitational force $F_G = G(\varepsilon_0\mu_0 h)^2 f_1 f_2/r^2$ is of electromagnetic nature, since ε_0 is the electric constant and μ_0 is the magnetic constant. So, the gravitational force is just like the electromagnetic force generated by virtual photons, where both the virtual photons and the gravitons propagate in the ZPF with the constant velocity c_0. This leads to the relativistic gravitational force, where $F'_G = G(\varepsilon_0\mu_0 h)^2 f_1 f_2/[r^2(1 - v^2/c_0^2)] = Gm_1m_2/[r^2(1 - v^2/c_0^2)]$ has the same structure as the relativistic equation for the electromagnetic force. This relativistic gravitational force can, for example, be verified by its ability to explain the apsidal precession of the planet Mercury, [194] and the other relativistic phenomena that involve the gravitational force.

4.1.2 Relativity has a Physical Explanation

Let us consider two particles A and B, which move to the right with the common velocity v relative to the ZPF and thus relative to the propagation velocity of the gravitational force. Let A be to the left of B. As the force decreases with the square of the distance, and since B moves away from A during the time it takes the force to get from A to B, the force on B will decrease; and since A approaches B during the time it takes the force to get from B to A, the force on A will increase. Because the shortest distance (from B to A) gives the greatest force, the body will shrink. This is due to obvious physical conditions. And likewise, since the physical dimensions of a moving clock are subjected to a length-contraction, the clock time, depending on the physical construction of the clock, will change. But the length-contraction does not affect the passage of time.

From the constant propagation velocity of the forces (such as virtual photons and gravitons) relative to the ZPF, it is possible to calculate the relativistic forces on bodies that move relative to the ZPF and thereby relative to the forces that bind the bodies together. The theory explains in this way all the relativistic relations such as the length contraction, the "time dilation", the relativistic mass, the mass-energy equivalence, the black holes, the influence of the gravitational field on the clock-time, and the deflection of matter and energy in a gravitational field, since the mass m of wave-particles, such as light, is equal to $m = \varepsilon_0\mu_0 E = \varepsilon_0\mu_0 h \cdot f$ (see eq. 2.43). The QET gives, in this way, a physical explanation of relativity.

Length contractions are normally linked to the Theory of Special Relativity, which can be deduced from the following two postulates (ref. [195], p. 1):

(1) *Principle of relativity (Galileo)*: "No experiment can measure the absolute velocity of an observer; the results of any experiment performed by an observer do not depend on his speed relative to other observers who are not involved in the experiment."

(2) *Universality of the speed of light (Einstein)*: "The speed of light relative to any unaccelerated observer is $c_0 = 3 \times 10^8$ m/s, regardless of the motion of the light's source relative to the observer. Let us be quite clear about this postulate's meaning: two different unaccelerated observers measuring the speed of the same photon will each find it to be moving at 3×10^8 m/s relative to themselves, regardless of their state of motion relative to each other."

AD (1): In chapter 2.2.4, it was mentioned that S. Marinov had measured the absolute speed of the earth (or rather the speed of the experimental setup) relative to the

ZPF to 362 ± 40 km/s, [5] and at the end of the same chapter, it was likewise mentioned that Wilkinson Microwave Anisotropy Probe (WMAP) had found that our local group of galaxies is moving at 369 ± 0.9 km/s relative to the cosmic microwave radiation. [6] The two measurements clearly show that the "principle of relativity" is wrong.

AD (2): If the velocity of the electromagnetic field (such as the light and the electromagnetic force) had the same velocity relative to any observer/object regardless of its state of motion, no length contractions would arise. This association between the extent of a body and its velocity relative to the ZPF has also been found by H. A. Lorentz, [2] and it has been found here by the use of Coulomb's law, chapter 2.3.1. So "the universality of the speed of light" is also wrong.

Both special relativity and general relativity are founded on these two postulates.

4.1.3 Inertial Systems and Relativity

In relation to relativistic issues, inertial systems ought to be depicted symmetrically, as they otherwise give a distorted picture of reality (Fig. 4.1).

Consider the interrelations between two coordinate systems that move relative to each other in the ZPF. Let us regard the two coordinate systems S and S', and assume that the length of the coordinates in S is equal to the length of the coordinates in S', when both the coordinate systems are at rest in relation to the ZPF, so $|x| = |x'|$, $|y| = |y'|$, and $|z| = |z'|$, when the coordinates x, y, z in S, and x', y', z' in S', all have a velocity equal to zero relative to the ZPF.

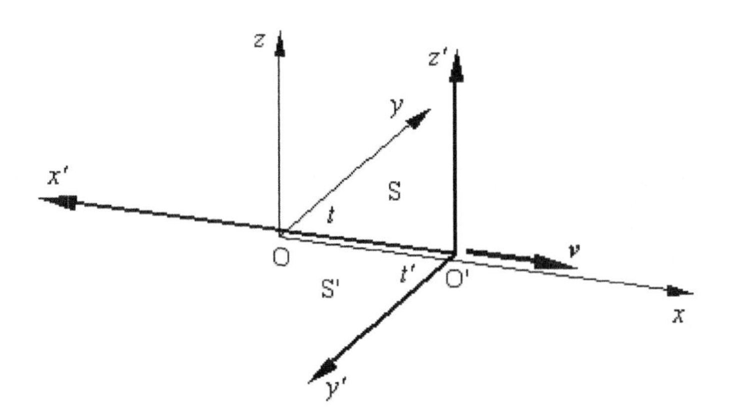

Figure 4.1: The length contraction of the coordinate axes.

We now assume that the coordinate system S' is given a constant velocity v relative to the coordinate system S, such that the coordinate system S' has a parallel motion along the x-axis, and the origin O' moves out of the positive x-axis in the reference system S, which we assume is at rest in the ZPF. Furthermore, the two reference systems are arranged so they are fully symmetrical.

According to the QET the relativistic relations will have the following explanation.

Seen from the stationary system S, the x'-coordinate will, because of the motion of the x'-axis relative to the propagation velocity of the electromagnetic forces that hold the axis together, be equal to $x\sqrt{1 - v^2/c_0^2}$, measured in the system S. This means that seen from S, the length of the x'-axis will be exposed to a length contraction, because of its velocity v relative to the ZPF, so the length of the coordinate axis from 0 to x' measured in the coordinate system S becomes:

$$|x'| = |x|\sqrt{1 - v^2/c_0^2}.$$

Seen from the moving system S', the x-coordinate will, because the x-axis is at rest relative to the ZPF, be equal to $x'/\sqrt{1 - v^2/c_0^2}$, measured in the system S'. This is because of the length contraction of the x'-axis, owing to its velocity v relative to the propagation velocity of the electromagnetic forces that hold the coordinate axis together. This means that seen from S', the x-axis will be exposed to a length dilation, because it is at rest relative to the ZPF, so the length of the x-axis from 0 to x measured in the coordinate system S' becomes:

$$|x| = |x'|/\sqrt{1 - v^2/c_0^2}.$$

It is seen that there are no paradoxes (read flaws); as both observers find exactly the same value for the length contraction.

The y'-axis and z'-axis, which both are perpendicular to the direction of motion, will however, only be exposed to minor length contractions. This is because the speed of a solid body relative to the ZPF has an influence on the relative velocity of the electromagnetic forces that hold the body together. If we take a look at the drawing in ch. 2.3.2, it is easy to imagine that all the oblique forces (not shown in the figure) between the different atoms, will be exposed to changes, depending on their angle in relation to the direction of motion. So a solid body that moves in the x-direction will also be exposed to length contractions in the y- and z-direction. This correlation between the extent of a body and its velocity relative to the ZPF was also found by H. A. Lorentz. [2]

Thus: relativity arises as a consequence of the final constant velocity of the forces relative to the ZPF, such that when a body moves relative to the ZPF, it also moves relative to the propagation velocity of the forces, whether it is the gravitational field or the electromagnetic field. Relativity is, accordingly, a physical quality.

But according to the Theory of Special Relativity the length contraction in system S' seen from S, will be equal to the length contraction in system S seen from S', since they both move with the same velocity relative to each other. Thus two identical length contractions arise, one in S and one in S' - so it will therefore not be possible to register any physical length contractions.

4.1.4 Michelson-Morley's Experiment can only be explained by a Length Contraction

The Michelson-Morley experiment was an attempt to examine whether an ether actually existed. The experiment compared the speed of light in perpendicular directions, to detect the relative motion of the light through the stationary ether.

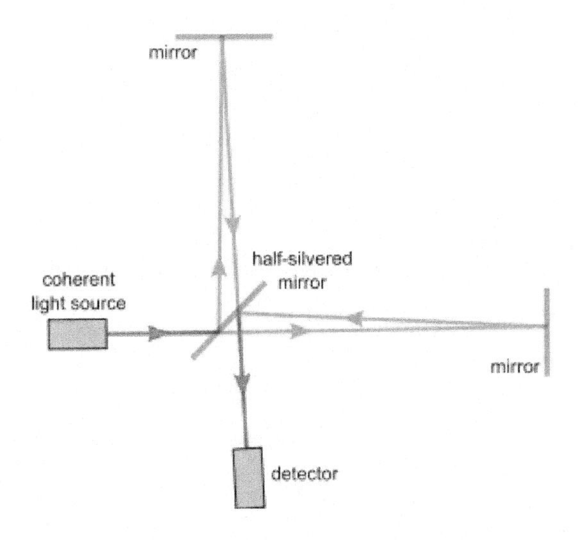

Figure 4.2: The Michelson and Morley experiment.

The experiment was an interference experiment, which, because of the changing velocities in relation to the ether, should have resulted in a change of the interference patterns, when the experimental setup followed the movements of the Earth around the Sun - but the experiment showed no changes in the interferences pattern. That is to say, Michelson and Morley found no difference between the speed of light in the direction of movement, and the speed of light at right angles.

There are two possible solutions to the Michelson-Morley experiment: 1) The light follows the movement of the experimental setup all the way around the Sun, so the photons always have the same velocity relative to the experimental setup. 2) The length of the experimental setup changes according to its velocity relative to the ZPF.

If it is possible for the light to follow the experimental setup all the way around the Sun, so the photons have the same constant velocity relative to the experimental setup independent of its velocity, it could be an obvious solution. However, in order to examine the validity of the second solution, we will have to calculate the time it takes for the light to pass through the experimental setup, when the light moves through the ether.

Assume that the apparatus is moving in the x-direction with the velocity v in the ZPF and that the speed of light c_0 is constant in relation to the field. When the apparatus is at rest relative to the ZPF, we have $L_x = L_y = L$.

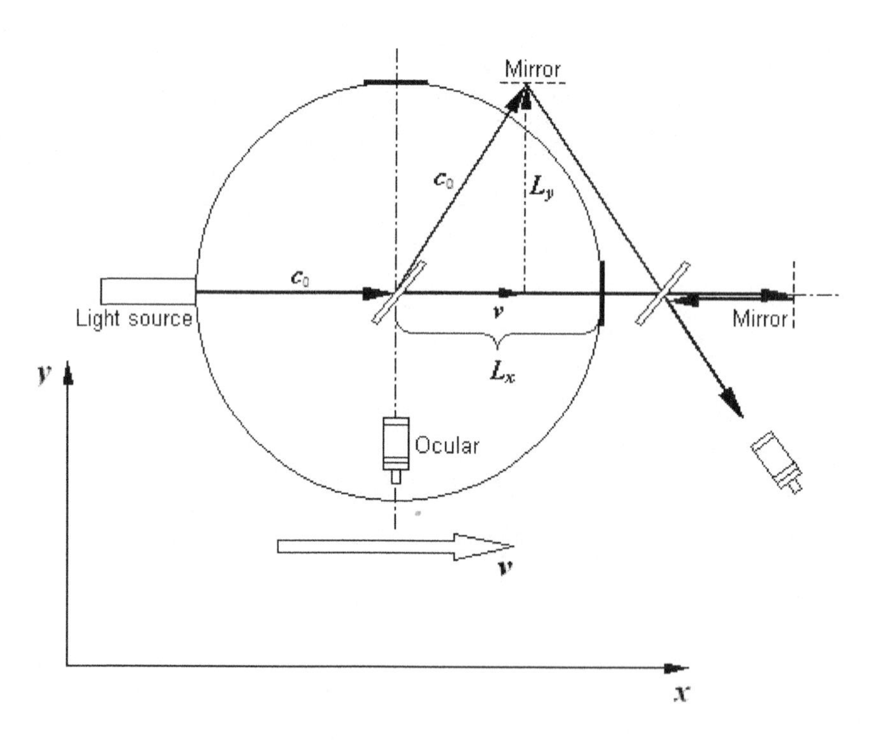

Figure 4.3: Michelsons-Morley's experiment.

First, we calculate the time it takes for the light to move from the semi-transparent mirror to the mirror in the y-direction. If t is the time the light needs to hit the mirror in the y-direction, then:

$$c_0^2 \cdot t^2 = L_y^2 + v^2 \cdot t^2$$

$$\Rightarrow (c_0^2 - v^2) \cdot t^2 = L_y^2$$

$$\Rightarrow t^2 = \frac{L_y^2}{c_0^2 - v^2}$$

$$\Rightarrow t = \frac{L_y}{(c_0^2 - v^2)^{1/2}}$$

The time it takes for the light to hit the mirror in the y-direction and return is thus:

$$\Rightarrow 2t = \frac{2L_y}{(c_0^2 - v^2)^{1/2}}.$$

The light in the x-direction needs the time t_1 to move from the semi-transparent mirror and hit the mirror in the x-direction, and the time t_2 to return to the semi-transparent mirror:

$$c_0 \cdot t_1 = L_x + v \cdot t_1 \Rightarrow t_1 = \frac{L_x}{c_0 - v},$$

and

$$c_0 \cdot t_2 = L_x - v \cdot t_2 \Rightarrow t_2 = \frac{L_x}{c_0 + v}.$$

The time it takes to complete a total movement in the x-direction then becomes:

$$t_1 + t_2 = \frac{L_x(c_0 + v) + L_x(c_0 - v)}{(c_0 + v) \cdot (c_0 - v)},$$

from which:

$$t_1 + t_2 = \frac{L_x c_0}{c_0^2 - v^2}.$$

According to the Michelson-Morley experiment, the time it takes to run through the two routes is identical, so $2t = t_1 + t_2$:

$$2t = t_1 + t_2$$

$$\Rightarrow \frac{2L_y}{(c_0^2 - v^2)^{1/2}} = \frac{2L_x c_0}{c_0^2 - v^2}$$

$$\Rightarrow L_y(c_0^2 - v^2) = L_x c_0 (c_0^2 - v^2)^{1/2}$$

Since the left and right sides of the equation are different and since the velocity v and the speed of light c_0 are both constant in relation to the ZPF, the difference must be due to a difference in the length between L_x and L_y. Let us denote the difference by γ (gamma), so $\gamma \cdot L_x = L_y$ or $L_x = 1/\gamma \cdot L_y$. In this way, we find that:

$$L_y(c_0^2 - v^2) = 1/\gamma \cdot L_y \cdot c_0(c_0^2 - v^2)^{1/2}$$

$$\Rightarrow \gamma = c_0(c_0^2 - v^2)^{1/2}/(c_0^2 - v^2)$$

$$\Rightarrow \gamma = c_0(c_0^2 - v^2)^{1/2}/[(c_0^2 - v^2)^{1/2}]^2 = c_0/(c_0^2 - v^2)^{1/2}$$

$$\Rightarrow \gamma = \frac{1}{\sqrt{1 - v^2/c_0^2}}. \tag{4.1}$$

We have thus found the Lorentz Factor: $\gamma = 1/\sqrt{1 - v^2/c_0^2}$.

As the experimental setup moves in relation to the electromagnetic field holding the equipment together, the experiment gives a physical explanation of the length contraction.

The apparatus consists of atoms held together by electromagnetic forces. When the apparatus is moving relative to the ZPF, the light and the electromagnetic force have exactly the same velocity relative to the apparatus, and since the apparatus because of its velocity relative to the electromagnetic field shrinks with the factor $(1 - v^2/c_0^2)^{1/2}$ in the direction of motion, no changes arise in the interference pattern during the motion of the apparatus around the Sun.

However, if we assume that the speed of light (c_0) is constant in relation to any object, regardless of the velocity of the object, the length contraction of the object cannot be explained. This is because, the speed of light as well as the speed of the electromagnetic field holding the object together, will in that case always be constant in relation to the object.

4.1.5 Test of the Quantum Ether Theory

In connection with the QET it would be obvious to perform the following tests:

- Try to find the gravitons by high-energy physics.
- Test whether the Plasma Redshift can explain the observed cosmological redshift.

4.2 Evaluation of the Euclidean Cosmos Theory

The evaluation compares the results of the theory with many of the observations and measurements already available of our Universe. It is done because, among other aspects, a theory must be judged on how well it describes the physical reality that surrounds us.

By comparing the ECT with the recent observations of our Universe, there have not been any observations - such as the early star formation, the network structure, the apparent expansion, the dark matter, the dark energy, etc. - which the ECT has not been able to explain in a logical manner.

Many of the physical phenomena relating to cosmology are of such a magnitude that it can be difficult to develop a test which is able to confirm or refute a theory about the Cosmos. In such cases, it may be more useful to regard the physical realities that the theory describes. Moreover, our Universe itself performs the most spectacular tests, whose results are hard to ignore, although laboratory experiments are pointing in another direction. Therefore, we will here stick to the physical observations in connection with the evaluation of the theory.

4.2.1 Free Particles can escape from a Black Hole

The source of the most crucial regenerative processes is the AGN in the middle of the galaxies. As the black holes are the result of a mass concentration in the Euclidean space, the particles will be able to leave the black holes if they are able to achieve a speed greater than the speed of light.

Particles bound together by forces propagating at the speed of light will not be able to leave a black hole, because the mass of the particles approaches infinity when the velocity approaches the speed of light (eq. 2.31). However, free particles, which are not bound together by any forces, may achieve speeds greater than the speed of light, c_0. Free wave-particles such as quarks can, for example, reach speeds up to $\sqrt{2}c_0$ in the empty space (ch. 3.4.7), and up to $\sqrt{3}c_0$ in a strong gravitational field (ch. 3.4.7), so the hot, dense quark-gluon plasma will have no problem leaving a black hole.

4.2.2 The Farthest Objects are observed as Galaxies and Quasars

As even the brightest objects such as galaxies and quasars that are more than 13.39 billion light-years away are currently too far away to be observed, it will first be possible to study such objects when we develop more powerful telescopes. But if we one day will be able to observe the limit of our universe, we will probably observe an asymmetric distribution of matter and energy, reflecting our position in relation to the center of our Universe.

Since all the universes are closed and therefore do not emit any radiation, it will normally not be possible to test whether there are more universes, and since the space between the closed universes only contains barren objects, neither will we receive any other forms of information. Only in the cases when our universe merges with another universe will we be able to receive information from the outside world. However, there is nothing to prevent us from testing the part of the theory that relates to our own closed universe.

4.2.3 The Oldest Star is older than the Big Bang

One single star with an age greater than the time since the Big Bang is enough to overthrow the Big Bang theory, and thus indirectly supports the ECT. Researchers at the Space Telescope Science Institute in Baltimore have found a star with an age of 14.5 billion years plus-minus 0.8 billion years, [196] which with large probability makes it older than the calculated age of our Universe of 13.80 $\pm$ 0.04 billion years.

The star can be seen with binoculars. It is, to be precise, an apparent 7.223 magnitude [197] high-velocity Population II subgiant with a low metal content [Fe/H] = -2.40 $\pm$ 0.10, [198] about 190.1 light-years away from the Earth in the constellation Libra. [197], [199]

The sub-giant has a location in the Hertzsprung–Russell diagram (Fig. 4.4) where the absolute magnitude is most sensitive to stellar age, and because the sub-giant with spectral type A is bright, nearby, un-reddened and has a well-determined chemical composition, the age of the star can be determined with great precision. [196]

From the true distance of the star, an exact value for the star's intrinsic brightness can be calculated, by the help of which the age of the star can be estimated by applying theories about the star's burn rate, chemical abundances and internal structure.

The star, which is cataloged as HD 140283, has been known for more than a century. Its high rate of motion provides evidence that the star is a visitor to our stellar neighborhood, where its orbit carries it down through the plane of our galaxy, from the ancient halo of Population II stars that encircle the Milky Way, and eventually returns it to the galactic halo.

The star was probably born in a primeval dwarf galaxy, which was eventually gravitationally shredded and sucked in by the Milky Way more than 12 billion years ago. The star has since retained its elongated orbit from that event, which is why it now passes through the solar neighborhood at a speed of 1,300,000 km per hour. However, if the cradle of the star was a metal-poor dwarf galaxy, the dwarf galaxies, which appear to be coeval with globular clusters, are probably older still. [200]

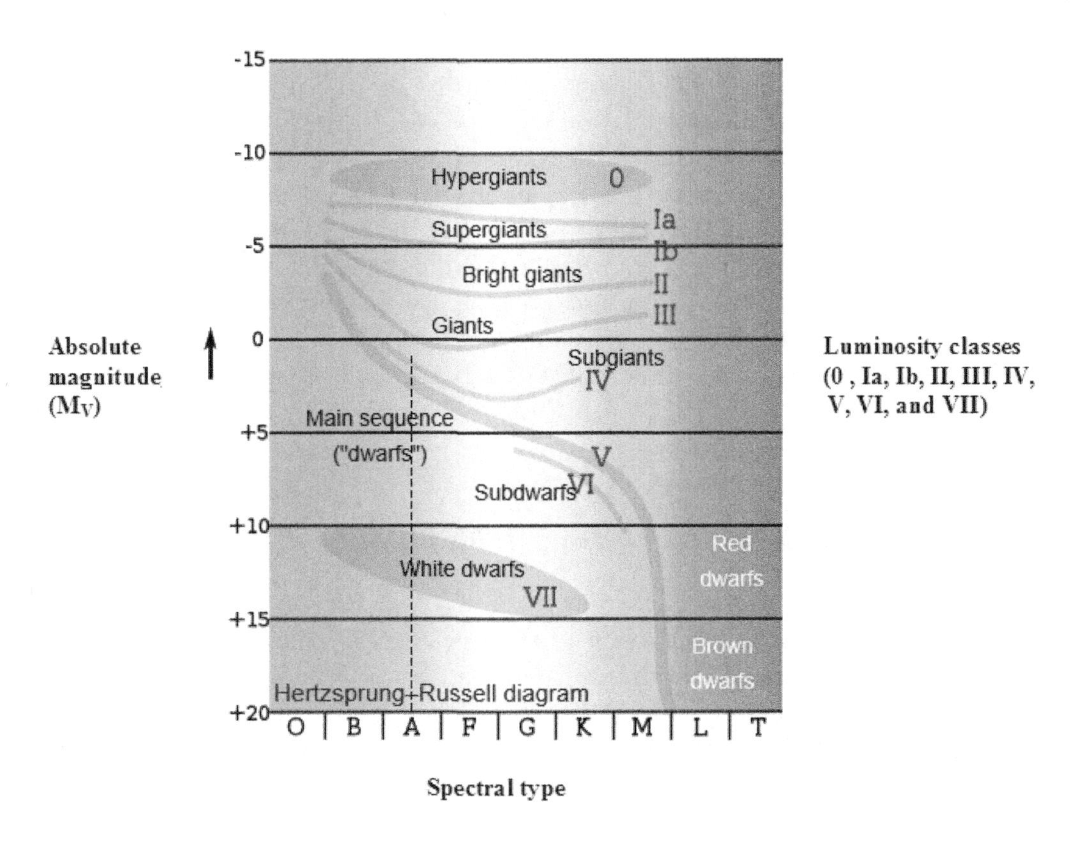

Figure 4.4: Hertzsprung-Russell diagram.

Because Population II stars collectively represent an older population of stars, the star has a deficiency of heavier elements compared to other stars in our galactic neighborhood, such as the Sun, which formed later in the disk. This means that the star was formed at a very early time before the molecular clouds were filled with heavier elements from supernova explosions. [199] Nonetheless, it contains still so much metal, that, at the time of its formation, there must have been enough to provide it with a certain metallicity.

The result is that the star's age is estimated to 14.5 ± 0.8 billion years, with an uncertainty, which makes it comparable with the age of the Universe at 13.799 ± 0.021 billion years. So when the time for the creation of the environment is included, which includes a condition for the generation of a star with such a metallicity, [196] we do have a problem.

The metal content of HD 140283 is equal to [Fe/H] = -2.40 ± 0.10. This is at least ten thousand times more than the first primordial gas clouds contain, which are estimated to have a metallicity of less than -6.0. The question is, where does this metal come from, and how long does it take to produce such a quantity of metals?

Despite HD 140283 being the oldest known star for which a reliable age has been determined, it is, given its low but non-zero metallicity, not quite a primordial star. It has been proposed that so-called Population III stars might exist, which could contribute to the metal content, but since HD140283 is already of the same age as the Universe, there has not been the time for such a Population III to arise and spread its metals through supernova explosions. If we add to the age of the star the time it takes to create the surroundings that are a prerequisite for the found metallicity plus the birth and development of the star, we have a universe with an age that far exceeds the time since the Big Bang. There is therefore only one explanation for the existence of such a star, it must have been present at the time of the Big Bang. The existence of the star HD 140283 supports in this way the Euclidean Cosmos Theory.

4.2.4 Quasars are observed between 750 million and 13.36 billion LY away

As our Universe as we know it according to the Euclidean Cosmos Theory has existed since the last collision with another closed universe, the activity of the stars, galaxies and quasars must be reasonably constant, when we look back in time. And as quasars or quasi-stellar radio sources, because of their luminosity, are the most energetic and distant members of the class of objects called AGNs, they are also among the best sources to use for looking back in time.

Quasars were first identified as being high redshift sources of electromagnetic energy, including radio waves and visible light. In size, they first appeared to be similar to stars, however with the development of new types of telescopes the quasars looked rather like extended sources similar to galaxies, [201] and since their spectra, due to multiple scattering, contain very broad emission lines unlike any known from stars, they were given the name "quasi-stellar".

A quasar is a compact region in the center of a massive galaxy that has a central SMBH at the center, [202] with a maximum density as a compact neutron star, and where the size of the galaxies is around 10-10,000 times the Schwarzschild radius of the black hole. When combined with the plasma redshifts, the implications of the very high redshifts are that quasars can be very distant and ancient. [203] Because they, apart from some luminous galaxies, are among the most luminous, powerful and energetic objects in our Universe, they are also among the farthest visible objects.

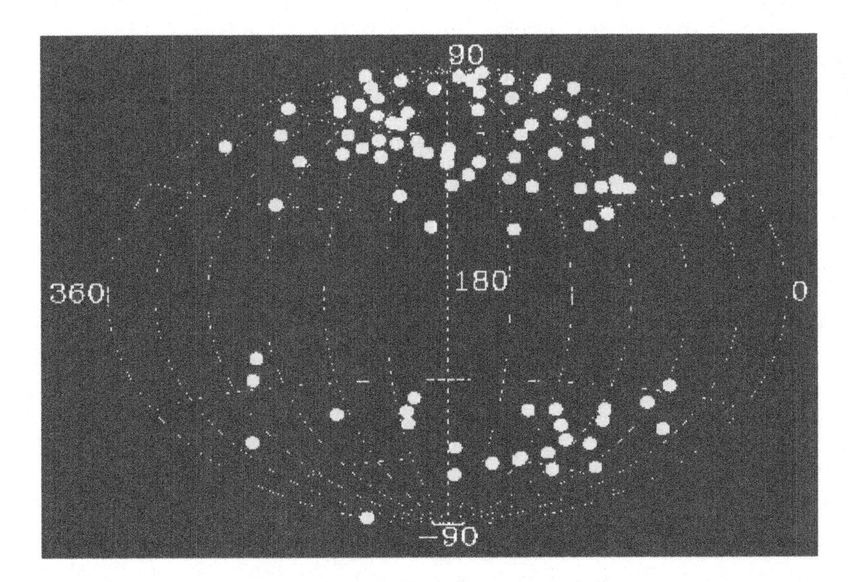

Figure 4.5: An all-sky image of the distribution of some of the brightest 86 quasars.

Each of the quasars emits up to a thousand times the energy output of the Milky Way, which contains about 300 billion stars. This radiation is emitted almost uniformly across the electromagnetic spectrum, from X-rays to the far infrared. Most quasars emit the near-ultraviolet wavelength of the 121.6 nm Lyman-alpha emission line of hydrogen, but due to the tremendous redshift of these sources, the peak luminosity has been observed as far to the red as 900.0 nm, in the near infrared.

Quasars are powered by accretion of material onto SMBHs at the center of the galaxies. Some material may fall directly onto the black hole, while other material may have an angular momentum around the black hole, which causes matter to collect in an accretion disc. [204] Because some quasars display very rapid changes in luminosity, which are fast in the optical range and even faster in the X-ray range, they define an upper limit to the volume of the core of the quasars; which is not much larger than our Solar System. [205]

The emission of such large amounts of power from such a small region requires a power source far more efficient than the nuclear fusion that powers the stars. This calls for an astonishingly high energy density, which supports that energy is generated by a black hole with a neutron star at the center. It has been found that each neutron can convert around 99% of its mass into pure energy (ch. 3.4.3), while the proton-proton chain of the nuclear fusion process that dominates the energy production in Sun-like stars can convert around 0.7%, [206] and if it was not possible for matter and energy to leave the black hole, the galaxies would long ago have ended up as huge black holes.

The masses of the large central black holes in quasars have been measured to $10^6 - 10^9$ solar masses. Several dozen of our nearby large galaxies, which show no signs of activity, have been shown to contain similar central black holes in their nuclei, so it is thought that all large galaxies have a black hole at the center, but only a small fraction are active and therefore seen as quasars. The quasars show us the locations where massive black holes grow in step with the mass of stars in their host galaxy.

It seems as if the quasars were much more common in the past, it is however because they, in addition to some galaxies, are energetic enough to be observed. This energy production ends when the SMBH has consumed all the gas and dust in its vicinity. This means that it is likely that the galaxies, including our own Milky Way, periodically undergo active phases, and settle down whenever they have exhausted the supply of material to feed their central black holes.

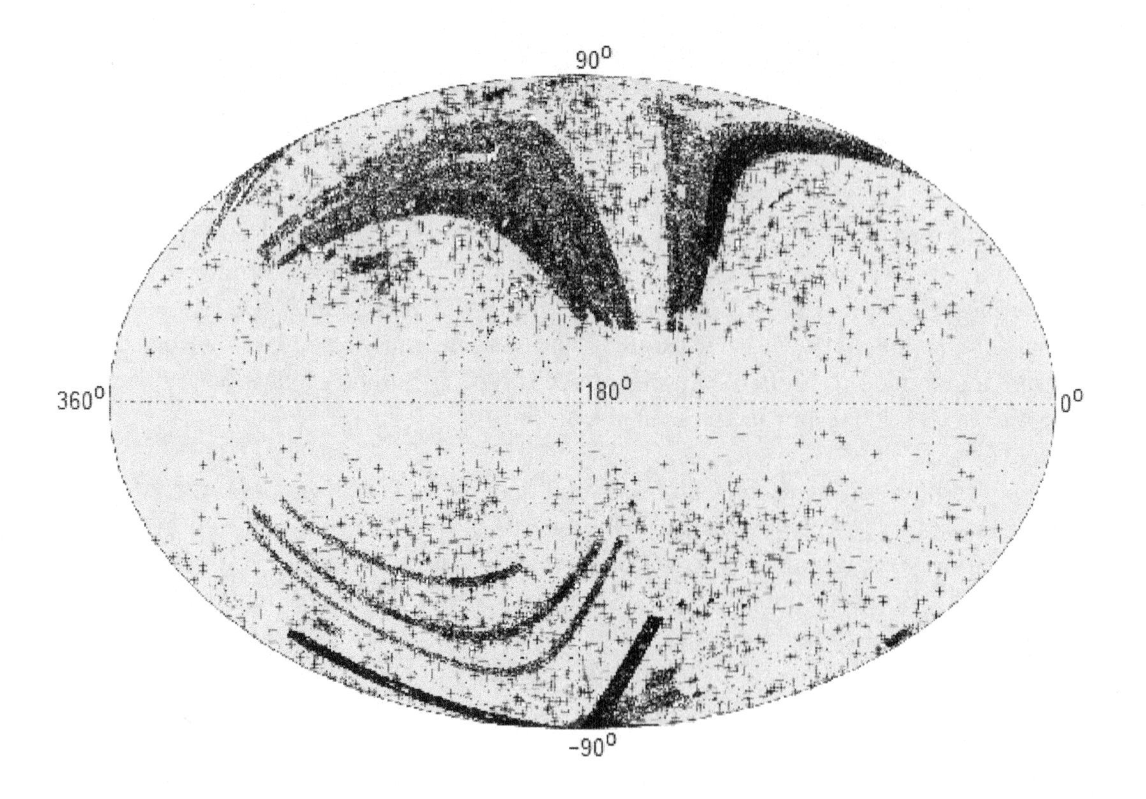

Figure 4.6: Distribution of Quasars and Active Galactic Nuclei in the sky. Small dots represent quasars brighter than V < 17 mag, while crosses represent quasars brighter than V − 17 mag. [209]

In Fig. 4.6 the dots and crosses represent the positions of quasars discovered to date, taken from the Veron-Cetty & Veron Catalog of Quasars and Active Galactic Nuclei, 2006. The gap in the distribution between the northern and southern galactic hemispheres is due to the shade of the galactic plane. According to the distribution of the quasars in the figure, it can be seen that there is an overall uneven distribution of quasars. This picture recurs to some extent, when we consider Fig. 4.8.

More than 300,000 quasars are known, [207] and the observed quasar spectra have redshifts between 0.056 and 7.085. Applying Hubble's law to these redshifts shows that they are between 750 million and 13.36 billion light-years away, [208] where the latter is at the very edge of our observable Universe.

The brightest known quasars devour 1000 solar masses of material every year, and the largest known quasar is estimated to consume matter equivalent to 600 Earths per minute, which is the ultimate verification of the reuse of energy. Since it is difficult to deliver fuel to quasars for billions of years, quasars become ordinary galaxies when they have finished consuming the surrounding gas and dust. Quasars may later be ignited or re-ignited in dormant galaxies. This can be due to different scenarios, such as when they later merge with other galaxies, so the black holes are infused with fresh sources of matter. It has in fact been suggested that a quasar could form when the Andromeda Galaxy collides with our own galaxy in approximately 4.5 billion years. [204], [210]

The existence of gigantic black holes as far as we can see, and the existence of resources that can deliver such huge amounts of matter that they can feed the black holes with an almost continuous flow of energy, can only be explained by that the Universe is much older than the Big Bang Theory prescribes.

4.2.5 The Old Quasars contain Metals and Carbon

The most common gases in the Universe are hydrogen and helium, so within astronomy, metallicity, which is designated Z, is the fraction of matter that is made up of chemical elements other than hydrogen and helium.

Through spectroscopic observations it has been demonstrated that there is a metal enrichment of the gas surrounding the nucleus of the galaxies, even for the highest redshift quasars. [211] It has for instance been found that even quasars, at a redshift $z \sim 6$, which corresponds to an age of 13.2 billion years, have a metallicity several times the metallicity of the Sun, [211] and in large galactic halos, at high redshifts, the star formation rate turns out to be very high, yielding a quick increase of the metallicity.

Several direct chemical measurements of circumnuclear gas in high-z quasars have established that both star formation and metal enrichment are fastest in the central regions. If we look at the density of the innermost 10% of the galaxy mass, the metallicity of the gas is between 3 and 5 times the solar metallicity, with a general trend to increase with mass. [212]

Furthermore, statistical studies of local elliptical and spiral galaxies hosting SMBHs, show that the most massive galaxies are also the most metal-rich, the reddest and the oldest. [213], [214] These galaxies show an excess of elements such as O, Ne, Mg, Si, S, Ar, Ca, and Ti, compared to the Sun. [214] This is suggestive of a very intense and short star formation activity or very old galaxies, as the bulk of the star formation must be completed before supernovae can contribute to the abundance of metals in the interstellar medium.

By using near infrared and optical spectra, the metallicity of a sample of quasars in the redshift range $4 < z < 6.4$ has been investigated, where the redshift range $4 < z < 6.4$ corresponds to an age of 12.7 Gyr to 13.3 Gyr. On average, the observed metallicity changes neither among quasars in the observed redshift range $4 < z < 6.4$, nor when compared with quasars at lower redshifts.

Despite the very high metallicity in quasars at $z \sim 6$, at an age of 13.2 Gyr, there is an apparent lack of evolution in the metal and carbon content among quasars, at $z < 6$, and the minimum enrichment timescale of carbon is about 1 Gyr, i.e. longer than the age of the Universe at $z \sim 6$, which is only about 0.57 Gyr. [211]

Seen from the Big Bang theory this is certainly puzzling, since, according to this model, the Universe starts from scratch at time zero. It might therefore be expected that there was a steady increase of the metal content, because of the steady supply of metals from the series of supernova explosions, which continuously provide the Universe with heavy particles.

Given that, according to the Euclidean Cosmos Theory, the Universe has always existed, the distribution of the metallicity depends only on the location inside the Universe, where density and activity are largest at the center. But because we only are able to observe a smaller section of our Universe, we can assume that the metallicity is constant inside this area. However, since we are only able to observe the most massive galaxies and quasars at the limit of our observable universe, it will only be the galaxies and quasars with the highest activity, and thereby the highest metallicity, we find at these distances.

4.2.6 The Dark Matter Halos consist of Baryonic Matter

From the rotation curves of galaxies, which can be rendered as graphical plots of the orbital speeds of the visible stars versus their radial distance from the center of the respective galaxies, is has been found that the rotational speeds of the stars are almost the same or increase with distance. [215] This is in contrast to the orbital velocities of planets in planetary systems, where the velocities are declining with distance. The velocities reflect in both cases the mass distribution within the systems, so in order to account for the observed speeds in the galaxies, it will be reasonable to assume the existence of "dark matter" which encloses the galaxy. Accordingly, the presence of dark matter in the halo can be derived from the gravitational effect of the halo on the rotational curve of the spiral galaxy. Without large amounts of mass everywhere in an approximately spherical halo, the rotational speed of the galaxy would fall at great distances from the center of the galaxy.

Observations of spiral galaxies, and especially radio observations of line emissions of neutral atomic hydrogen, show that the rotational speeds do not fall with the distance from the galactic center. The cold atomic hydrogen gas that resides in the large spiral galaxies is gravitationally confined to the galactic disk and extends much further out than the visible stars, sometimes up to five times the diameter of the visible spiral. This gas can be mapped by radio telescopes. The indirect detection of dark baryonic matter as well as dwarf galaxies has an influence on planetary dynamics and the galaxy evolution, which is dominated by such satellites. The dark satellite galaxies create disturbances in the cold atomic hydrogen gas at the edges of the spiral galaxy's disk, and these perturbations reveal the mass, distance and location of the satellites. It has been observed that the outer edges of the Milky Way appear to be orbited by hundreds of satellite galaxies. However, while a few satellites are visible, like the Magellanic Clouds, many other galaxies are too dim to see. [216]

Overall it is found that the visible disk of the Milky Way Galaxy is embedded in a much larger roughly spherical halo of dark matter, whose density drops off with the distance from the galactic center. It is estimated that about 95% of the galaxy is composed of dark matter that only interacts with the rest of the galaxy through gravity. The luminous matter of our galaxy consists of approximately 9×10^{10} solar masses, while the dark matter is likely to include between 6×10^{11} to 3×10^{12} solar masses. [217] From the life cycle of stars (Fig. 3.8), it can be seen that the hydrogen and helium from the stellar nebulae eventually end up as black holes, and since the Cosmos has existed for an infinitely long time according to the Euclidean Cosmos Theory, and since it only requires three solar masses to create a black hole, the universe will be teeming with such objects, where the black holes in the outer part of the galaxy make up the galactic halo that surrounds the galactic disk and stretches far beyond the edge of the visible gaseous galaxy, while the huge black holes at the center of the galaxies act as AGNs, which through regenerative processes supply the nebulae with hydrogen and helium.

Through observations, we know that AGNs consist of SMBHs in the middle of a massive galaxy. [202] When a SMBH has accumulated so much material that it generates a regenerative process; the jet of photons, nucleons and leptons will be spewed out into the surrounding galaxy and the universe, where the particles may reach velocities ranging from $0.1 \cdot c_0$ to $0.4 \cdot c_0$. [218] The nucleons and leptons that remain in the galaxy will gravitate towards the largest mass concentrations, where they will create the nebulae that supply the energy for the next generations of stars; [219] while the rest ends up in the intergalactic medium as cosmic rays.

It has been observed that massive black holes are present in essentially all the local galaxies with a substantial spheroidal component, [220] and the correlation between the mass of the black hole and the stellar velocity dispersion is found at redshift up to 3, corresponding to 12 Gly. [221] The accretion rate onto the central black hole is observed to be directly proportional to the luminosity produced by the AGN, [222] where the black hole nucleus of the galaxy that makes up the AGN will shine as an optical quasar or Seyfert until the reservoir of mass is exhausted. From this point on the host galaxy evolves passively and the black hole becomes dormant until it is reactivated by accreation of matter. [221]

A map of interstellar hydrogen could also answer how the galaxies manage their gas supply. [223] According to observations, most galaxies have a reservoir of gas that is large enough to make stars for another billion years or so, and during that time they are likely to get new supplies of leptons, protons, and alpha particles from the AGNs.

If black holes are to constitute the dark matter, they must be the end result of the life cycle of energy in the universe and meet the requirement, that they do not occupy so much space that they hinder the transparency of the universes. The Schwarzschild equation (eq. 2.48) gives the following connection between the Schwarzschild radius r_S and the mass M_{bh} of a black hole:

$$r_S = \frac{2GM_{bh}}{c_0^2},$$

where $G = 6.67 \times 10^{-11}$ m^{3}kg^{-1}s^{-2} is the gravitational constant, M_{bh} is the total mass of the black hole, and $c_0 = 3 \times 10^8$ m/s is the speed of light.

From earlier (ch. 3.2.3) we know that a neutron star has a mass of between 1.1 $M_\odot$ and 3.2 $M_\odot$, and from chapter 3.4.1 we know that a neutron star needs a mass greater than about 3.2 solar masses in order to exceed the Tolman-Oppenheimer-Volkoff limit. When it exceeds this limit it will generate an enormous explosion, so the neutron stars must lie in the interval between 1.1 $M_\odot$ and 3.2 $M_\odot$, where they constitute the center of dark stars and black holes.

When a neutron star at the center of a black hole explodes due to accretion of matter - which generates the necessary external pressure on the neutron star so it reaches the Tolman-Oppenheimer-Volkoff limit - it can be observed as a giant stellar explosion called a hyper-nova. Such explosions have been observed as energy bursts at enormous velocities and flashes of gamma rays in the form of jets.

To make an estimation of the relation between the black holes and neutron stars, we compare the density of the black holes with the density of the neutron stars, when the neutron stars approach a mass equal to 3.2 $M_\odot$.

Since the solar mass equals $M_\odot = 1.99 \times 10^{30}$ kg, the Schwarzschild radius of a black hole with a mass equal to 3.2 $M_\odot$ becomes:

$$r_{S\ at\ 3.2M_\odot} = \frac{2G \cdot 3.2M_\odot}{c_0^2} = \frac{2 \cdot 6.67 \times 10^{-11} \cdot 3.2 \cdot 1.99 \times 10^{30}}{(3 \times 10^8)^2} \approx 9,400 \text{ m}.$$

The density ρ_{bh} of a black hole with the mass $M_{bh} = 3.2M_\odot$, must be equal to the mass of the black hole divided with its volume $V_{bh} = 4\pi r_{bh}^3/3$, so:

$$\rho_{bh\ at\ 3.2M_\odot} = \frac{M_{bh}}{V_{bh}} = \frac{3.2M_\odot}{4\pi r_{bh}^3/3} = \frac{3.2 \cdot 1.99 \times 10^{30}}{4 \cdot 3.14 \cdot 9400^3/3} \approx 1.8 \times 10^{18} \text{ kg/m}^3$$

From chapter 3.4.6 it can be seen that the saturation density of a neutron star at the time of the explosion lies between (4 times 2.8×10^{17} kg/m^3) and (8 times 2.8×10^{17} kg/m^3), or between (1.3×10^{18} kg/m^3) and (2.2×10^{18} kg/m^3), so it likely that numerous black holes exist where the central neutron star has a mass less than the Tolman-Oppenheimer-Volkoff limit of 3.2 solar masses.

Let us evaluate whether it is possible for the black holes to make up dark matter without being noticed. To make an estimate of the volume of the black holes, we can examine how many black holes N_{bh} of 3.2 solar masses the volume of the Sun can contain. From the radius of the Sun $R_\odot = 695,700$ km and the found radius of a black hole of 3.2 solar masses, $r_{bh} \approx 9,400$ m, we can find the volume of the Sun $V_\odot$ and the volume of the black hole V_{bh}. The number of black holes, the volume of the Sun can contain, then becomes:

$$N_{bh} = \frac{V_\odot}{V_{bh}} = \frac{4/3\pi {R_\odot}^3}{4/3\pi {r_{bh}}^3} = \frac{{R_\odot}^3}{{r_{bh}}^3} = \frac{695,700^3}{9.4^3} = 4.1 \times 10^{14} \text{ black holes}$$

By comparison the Milky Way contains around 9×10^{10} stars; [217] and since the black holes are so small, they will almost be impossible to detect when they are distributed in and around our galaxy, but their presence can be inferred by their gravitational pull.

It may also happen that a neutron star at the center of a black hole explodes due to accretion of matter - which generates the necessary external pressure so the neutron star reaches the Tolman-Oppenheimer-Volkoff limit - this can be observed as a giant stellar explosion called a hyper-nova. Such explosions have been observed as energy bursts at enormous velocities and flashes of gamma rays in the form of jets.

Even if the black hole has become much larger than the neutron star in the center, the density of the neutron star does not necessarily achieve the required saturation density for an explosion. This is because the total density of the black hole decreases with size. So, although the density of the neutron star in the center of the black hole is very large, the density of the remaining black hole can be rather sparse.

Let us try to add additional mass to a barren neutron star. It may for instance be 7 solar masses, so the black hole in total consists of 10 solar masses. As the solar mass is equal to $M_{\odot} = 1.99 \times 10^{30}$ kg, the Schwarzschild radius of the black hole becomes:

$$r_{bh} = \frac{2G \cdot 10M_{\odot}}{c_0^2} = \frac{2 \cdot 6.67 \times 10^{-11} \cdot 10 \cdot 1.99 \times 10^{30}}{(3 \times 10^8)^2} \approx 29,600 \text{ m}.$$

The density of a black hole of 10 solar masses then becomes:

$$\rho_{bh} = \frac{M_{bh}}{V_{bh}} = \frac{10M_{\odot}}{4\pi r_{bh}^3/3} = \frac{10 \cdot 1.99 \times 10^{30}}{4 \cdot 3.14 \cdot 29600^3/3} \approx 1.8 \times 10^{17} \text{ kg/m}^3$$

That is to say, the overall density of the black hole has become 10 times less, while the radius of the black hole has become about 3 times larger. So, when a barren neutron stars accumulates further mass with a lower density, it will most likely end up as a large black hole - or explode as a hyper-nova or an AGN.

As our Universe has existed since its last collision with another universe, our Universe will be teeming with black holes and other dark matter, which only can be inferred from their gravitational effect on visible matter. Furthermore, as can be seen from the calculation, they will hardly take up any room. It is quite unlikely that there exist a kind of dark matter in the outer space that is not present in our environment or that we have not seen in our high energy particle physics experiments, thus dark matter will consist of normal baryonic matter like protons and neutrons and other elementary particles, which make up the black dwarfs, barren neutron stars and black holes.

The distribution of dark matter in a galaxy follows the physical laws of the distribution of rotating matter around a common center of mass, where the lighter hydrogen and helium gravitate toward the center, while the heavier dark matter like black holes, barren neutron stars and black dwarfs circle around the galaxy in many different orbits far beyond the edge of the visible galaxy. Consequently, we find that most of a universe consists of a structure of black holes where the densest regions such as AGNs generate the lightest elements such as hydrogen and helium, which accumulate in the most massive regions where they create the stars that illuminate the universes.

So, in each of the closed universes there arises an equilibrium between the energy production and energy consumption.

4.2.7 The many Regenerative Explosions are mirrored in the CMB

The CMB provides some of the best knowledge we have about the structure, content and history of the Universe. For many years hundreds of experiments have been conducted to measure and characterize the signatures of the cosmic microwave background radiation. NASA's Cosmic Background Explorer (COBE) satellite has detected and quantified the large-scale anisotropies, and a series of experiments have later quantified CMB anisotropies on smaller angular scales.

By the year 2000, the BOOMERanG experiment (Balloon Observations Of Millimetric Extragalactic Radiation ANd Geophysics) had measured the CMB of a part of the sky and reported that the highest power fluctuations occur at scales of approximately one degree. Together with other cosmological data, these results implied that the geometry of the Universe is flat.

In June 2001, NASA launched a CMB space mission called WMAP (Wilkinson Microwave Anisotropy Probe). The first results from this mission were detailed measurements of the angular power spectrum at a scale of less than one degree. This space mission provided very accurate measurements of the large-scale angular fluctuations in the CMB structures. Later, in 2009, the European Space Agency (ESA) Planck Surveyor

Figure 4.7: All-sky map of the Cosmic Microwave Background.

was launched. It has measured the CMB at a smaller scale than WMAP and in 2013 the European-led research team behind the Planck cosmology probe released the mission's all-sky map of the CMB.

The map suggests that the subtle fluctuations in temperature were imprinted on the deep sky. From the CMB data, it can be seen that our local group of galaxies appears to be moving at 369 ± 0.9 km/s relative to the reference frame of the CMB. (ref. [6], p. 231). This is in accordance with Stefan Marinov's measurement of the velocity of the experimental equipment relative to the ZPF, which he measured to 362 ± 40 km/s. [5] The motion results in an anisotropy of the data as the CMB appear to be slightly warmer in the direction of motion than in the opposite direction. The interpretation of this temperature variation is a simple velocity redshift and blueshift due to the motion relative to the CMB. According to the ECT, the rest frame of the CMB is identical with the rest frame of the ZPF.

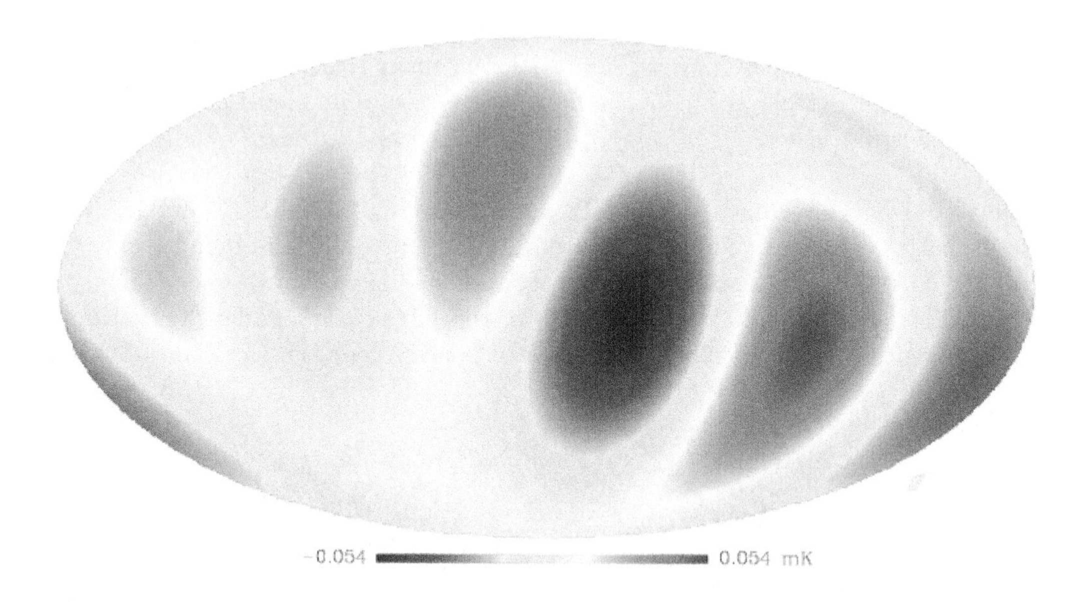

Figure 4.8: The lopsided Universe.

To measure the physical size of these anisotropies, it is necessary to turn the whole-sky map of temperature fluctuations into a power spectrum. The power spectrum encompasses fluctuations over the whole sky down to very small variations in temperature. Smaller details in the fluctuations tell us about the relative amounts of ordinary matter. However, some of the largest fluctuations - covering one-sixteenth, one-eighth, and even one-fourth of the sky - are bigger than any structure in the Universe, and therefore represent temperature variations across the whole sky. Recent observations with the Planck telescope, which is very much more sensitive than WMAP and has a larger angular resolution, confirm these observations. Since two different instruments recorded the same anomaly, instrumental error appears to be ruled out. [224]

The temperature variations are both larger than expected and aligned with each other to a very high degree. This is at odds with the standard cosmological model where the CMB anisotropies should be randomly oriented, not aligned. In fact, since the smaller-scale variations are random, it makes the deviation at larger scales much more remarkable. These large-scale deviations are reflected in temperature fluctuations much larger than any galaxy cluster. According to the ECT these large-scale deviations may reflect the position of the visible universe in relation to the center of our closed Universe.

If these large anomalies are due to primordial physical phenomena, the impact of the results is huge. In addition to specifying our location relative to the entire Universe, the fluctuations also indicate the network structure of the universe, such as the distribution of galaxies, galaxy-clusters, super-clusters, sheets, walls, and filaments, which are separated by immense voids, creating a vast foam-like structure called the cosmic web. These structures can only be created through hundreds of billions of years.

An evaluation of the Euclidean Cosmos Theory could include the following physical observations and experiments:

- Observe whether the formation of stars, galaxies, and quasars is reasonably constant during the observable lifetime of our Universe.
- Observe whether the age distribution of galaxies supports the theory.
- Experimentally verify the connection between the temperature of the intergalactic plasma and the plasma redshift.

Chapter 5

Conclusions

5.1 Conclusions

Based on positivistic, empirical science the book offers a thorough, logical insight and understanding of the physical world in which we live.

To gain an explanation of the structure and composition of the Cosmos, it has been necessary to establish two new theories: The Quantum Ether Theory and The Euclidean Cosmos Theory. The Quantum Ether Theory explains the physical questions regarding the structure and relativistic issues in connection with the Euclidean space, which is necessary to explain the questions in relation the composition of the Cosmos.

The theories are deduced from the most fundamental laws of classical physics. The only necessary deviation from classical physics has been a tiny alteration of Newton's law of universal gravitation, as it is assumed, that the gravitational force is transmitted by gravitons that propagate with the constant speed of light. This slight alteration of the theory has made it possible to deduce the relativistic relations in connection with the theory.

The book establishes that through experiments by Michelson and Morley, Stefan Marinov, Hendrik Casimir, Maxwell and the Wilkinson Microwave Anisotropy Probe, it has been proven that time is linear and universal, so that space and time can best be expressed by a three-dimensional space and a linear time. That is to say, the space is Euclidean, and since all the relativistic observations can be resolved by assuming that the gravitons propagate with the speed of light, a three-dimensional space with a linear time is the only realistic choice.

5.1.1 The Quantum Ether Theory

On the basis of: H. A. Lorentz's theory of space and time, the Casimir effect, Maxwell's equations, Michelson and Morley's experiment, Stefan Marinov's measurement of the absolute velocity relative to the electromagnetic field (to 362 ± 40 km/s), Wilkinson Microwave Anisotropy Probe's measurement of the absolute velocity of the probe relative to the reference frame of the CMB (to 369 ± 0.9 km/s) and the Quantum Field Theory, it can be concluded, that an ether of irreducible electromagnetic vacuum fluctuations at absolute zero temperature exists. This ether is called the zero-point field.

The infinite Euclidean Cosmos contains among other fields: the electromagnetic quantum field, the gravitational field and the quark gluon field, whose lowest possible energy states are named the zero-point field. These fields consist of virtual particles such as virtual photons, gravitons and gluons that establish the interaction between ordinary matter through an exchanges of virtual particles.

Beside the virtual particles the Cosmos contains ordinary matter in the form of wave-particles, which make up the quantum particles and the electromagnetic radiation, and James Clerk Maxwell showed that light is connected to the electromagnetic field by the following relation:

$$\varepsilon_0 \mu_0 = 1/c_0^2,$$

where ε_0 is the vacuum permittivity (called the electric constant) and μ_0 is the vacuum permeability (called the magnetic constant), so light moves with the constant speed $c_0 = 1/\sqrt{\varepsilon_0 \mu_0}$ relative to the ZPF. By the use of the mass-energy equivalence $E = mc_0^2$ and of Planck's radiation law $E = hf$, the mass of a wave-particle can be expressed by:

$$m = E/c_0^2 = \varepsilon_0 \mu_0 E = \varepsilon_0 \mu_0 h f.$$

From the expression it can be seen that the mass only depends on the frequency, since the other three factors ε_0, μ_0 and h, are constants. Since Planck's constant h establishes a quantization of the energy, radiation can only assume whole quanta with the energy $E = hf$, so it is obvious that the mass is quantized, and from the presence of the electric and magnetic constants we conclude that the mass is also of electromagnetic nature.

The gravitational force F_G between two wave-particles, with the masses $m_1 = \varepsilon_0 \mu_0 h f_1$ and $m_2 = \varepsilon_0 \mu_0 h f_2$, can according to Newton's law of universal gravitation be expressed by:

$$F_G = G m_1 m_2 / r^2 = G(\varepsilon_0 \mu_0 h)^2 f_1 f_2 / r^2,$$

where G is the gravitational constant, and f_1 and f_2 are the frequencies related to the two wave-particles. Since all the factors - except for the gravitational constant G and the distance between the masses r - describe quantized electromagnetic waves, the gravitational force is just like the electromagnetic force transmitted by virtual particles, which in connection with the gravitational field are called gravitons. Since the gravitons, like the virtual photons, are used in connection with quantized electromagnetic waves, they must be related to the virtual photons and, like all other massless virtual particles, propagate at the speed of light, c_0, relative to the zero field.

Normal baryonic matter is composed of protons, neutron and electrons, which constitute the atoms. The protons and neutrons make up the positive nucleus of an atom, while the electrons make up the negative electron cloud. So molecules and solids that consist of atoms are held together by electromagnetic bonds between the charged particles. Since the electromagnetic force is due to the electromagnetic field, it propagates with the speed of light c_0 relative to the zero-point field. So, when charged particles move relative to the electromagnetic field, the distance the forces have to travel between the particles will be altered, which creates a length contraction (ch. 2.3.1). According to Coulomb's law the force between two charges is equal to:

$$F_C = k_0 q_1 q_2 / r^2 \,, \tag{5.1}$$

where F_C is the force, r is the distance between two point charges q_1 and q_2, and $k_0 = 1/(4\pi\varepsilon_0)$ is a constant. When the two point charges are set in motion with the velocity v relative to the ZPF along the line of connection between the two charges q_1 and q_2, they will be exposed to a length contraction. Because of the difference in the length the forces have to cover from q_1 to q_2 and from q_2 to q_1, the force becomes equal to:

$$F'_C = k_0 q_1 q_2 / r'^2 = k_0 q_1 q_2 / [r^2(1 - v^2/c_0^2)], \tag{5.2}$$

where the relativistic length contraction $r' = r\sqrt{1 - v^2/c_0^2}$ is called the Lorentz contraction.

This argument can, of course, be repeated for the gravitational force $F_G = Gm_1 m_2/r^2 = G(\varepsilon_0\mu_0 h)^2 f_1 f_2/r^2$, so when two masses m_1 and m_2 are set in motion with the velocity v relative to the ZPF along the line of connection between the two charges m_1 and m_2, they will be subjected to a length contraction equal to:

$$F'_G = Gm_1 m_2/[r^2(1 - v^2/c_0^2)] = G(\varepsilon_0\mu_0 h)^2 f_1 f_2/[r^2(1 - v^2/c_0^2)]$$

This is the gravitational force, when the particles in the stationary system have the mutual spacing r and move along their connecting line with the velocity v.

It can be shown that the harmonic waves satisfy the Schrôdinger equation $\hat{H}\Psi(\vec{r},t) = i\hbar\frac{\delta}{\delta t}\Psi(\vec{r},t)$, as long as it applies that $c \cdot k = \omega$ or that the phase velocity $c = \lambda \cdot f$ (eq. 2.16). [31] The differential equation is a typical equation for the disturbance of a state of equilibrium and describes waves that propagate with the phase velocity c, which can be any velocity that satisfies the Schrödinger equation. By superposition of harmonic waves with the proper relationship between ω and k, it is therefore possible to form new solutions to the wave equation such as wave packets.

De Broglie assumed that Planck's radiation law $E = hf$ with the momentum $p = h/\lambda$ in the direction of propagation, where h is Planck's constant and λ is the wavelength, is true for all particles, so that atoms and molecules can be seen as waves with a frequency $f = E/h$ and a wavelength $\lambda = h/p$. De Broglie showed that the waves only form standing waves for certain discrete frequencies, corresponding to discrete energy levels.

As the particles move with their individual phase velocities, there is nothing to prevent the particles from moving faster than the velocity of light. When it occurs, the relativistic gravitational force will change sign, so it becomes repulsive.

Since the space is Euclidean even in connection with the black holes, where gravity becomes so strong that not even light can escape, particles such as quarks that are not bound together by gluons may act as free particles. It can be seen from the following, that such free particles are able to reach velocities up to between $\sqrt{2} \cdot c_0$ and $\sqrt{3} \cdot c_0$ in a central gravitational field, which is why the quarks are able to leave the black holes.

The total energy of any wave-particle must be equal to its rest energy relative to the zero-point field E_0 plus the energy that is added to the particle, such as kinetic energy E_k relative to the ZPF and potential energy U according to its position relative to external forces, so the total energy of a wave-particle is equal to:

$$E = E_0 + E_k + U,$$

where E_0 is the rest energy, E_k is the kinetic energy and U is the potential energy. For a mass with the rest energy $E_0 = m_0 c_0^2$, the kinetic energy $E_k = mv^2/2$ and the potential energy $U = -GMm/r$, the total energy $E = mc_0^2$ can be expressed by:

$$mc_0^2 = m_0 c_0^2 + mv^2/2 - GMm/r,$$

where G is equal to the gravitational constant and M is the total mass of the external mass distribution with the center of mass at the distance r from the wave-particle. So:

$$m = m_0/[1 - v^2/(2c_0^2) + GM/(rc_0^2)].$$

Since $E = mc_0^2$ and $E = fh$, we find that $fh = mc_0^2$ or $m = fh/c_0^2$ and $m_0 = f_0h/c_0^2$, where the zero indicates that the particle is at rest.

For a wave-particle with the rest energy f_0h, the kinetic energy E_k and the potential energy U the frequency f can be expressed by:

$$f = f_0/[1 - v^2/(2c_0^2) + GM/(rc_0^2).$$

The connection between the frequency of a stationary free wave-particle f_0 and the same wave-particle with the velocity v relative to the ZPF is:

$$f = f_0/[1 - v^2/(2c_0^2)],$$

which means that a free particle that is not affected by a gravitational field can obtain a velocity $v < \sqrt{2}c_0$, before its frequency becomes infinite, regardless of the wavelength of the wave-particle.

Therefore, we can conclude that it is physically possible for a free wave-particle to move at any velocity v less than $\sqrt{2}c_0$ and in the extreme cases, where the wave-particle is under the influence of a gravitational field from a neutron star, it can move at any velocity less than $\sqrt{3}c_0$ (ch. 3.4.7). Such velocities will, along with the high pressure and the explosive expansion, make it possible for a hot quark-gluon plasma to escape a black hole.

From Newton's law of universal gravitation, $F_G = GMm/r^2 = ma$, it can be seen that the acceleration $a = GM/r^2$ of the mass m is independent of the mass of the particle. That is to say, that both the force on the particle and the inertia of particle are proportional to the mass m, and thereby the frequency f of the particle. So all particles under the influence of a gravitational field will, irrespective of their mass, get the same acceleration, and thereby the same velocity. That both the force on a particle and the inertia of a particle are proportional to the "mass", we know from when we have a dog at a leash.

Hubble's law is the name of the astronomical observation that all objects observed in deep space are found to have a redshift proportional to their distance from the Earth. A redshift occurs when electromagnetic waves undergo a shift toward the less energetic end of the electromagnetic spectrum either due to the Doppler effect, a gravity displacement or a plasma redshift. The redshift leads to an increase in the wavelength of the electromagnetic radiation compared to the wavelength that originates from the source. This increase in wavelength corresponds to a decrease of the frequency.

When the photons lose energy during a plasma redshift, the photons transfer energy to the plasma in extremely small quanta. In the quiescent solar corona, this heating of the plasma starts in the transition zone to the solar corona and is a major fraction of the coronal heating. Generally, the part of the photon energy that is transferred to the plasma causes a significant heating of the plasma, so the plasma redshift contributes to the heating of the interstellar plasma, the solar corona, the corona of galaxies and the intergalactic plasma. In this way, the plasma redshift explains the solar redshifts and the redshifts of galactic coronas, and leads to a hot intergalactic plasma, which can explain the cosmological redshift, the microwave background and the X-ray background.

The assumption that the redshift could be interpreted as the velocity of the source, led to the idea that the speed of the receding galaxies should be proportional to their distance from the Earth, so the Hubble's law could be given the simple mathematical expression:

$$v_s = H_0 D_H,$$

where v_s is the speed of the source in agreement with the redshift, where H_0 is Hubble's constant and D_H is the distance from the source to the observer. But where should the energy come from to generate such an expansion?

In order to anchor the speed of light to the ZPF, the theory derives Maxwell's equations based on the properties of the ZPF. From the found wave function it can be seen that the electromagnetic field propagates with the constant velocity $c_0 = (\varepsilon_0 \mu_0)^{-1/2}$ in the ZPF, where (ε_0) is the vacuum permittivity (also called the electric constant) and (μ_0) is the vacuum permeability (called the magnetic constant) of the electromagnetic field. So the propagation velocity of electromagnetic waves, including light, has the constant speed c_0 relative to the ZPF. It means that light cannot follow Michelson and Morley's experimental setup.

On the basis of Planck's radiation law: $E = hf$ (where E is the energy, h is Planck's constant and f is the frequency of a wave-particle) and the mass-energy equivalence: $E = mc_0^2$, we find that the mass of a wave-particle can be expressed by: $m = hf/c_0^2 = \varepsilon_0 \mu_0 hf$. So, when we consider a wave-particle such as a photon with the frequency f, its mass $m = \varepsilon_0 \mu_0 hf$ has exactly the same properties as any other mass. Depending on the circumstances it may have a potential energy mgh, a kinetic energy $mc_0^2/2$ or a momentum $p = mc_0$, so mass and energy are two sides of the same coin. That is to say, Newton's laws can be used on wave-particles in the same way as if they were solid bodies.

Since light and all other wave-particles all have a mass equal to $m = \varepsilon_0 \mu_0 hf$, quantum waves will be attracted by a gravitational field. This leads to the bending of light and thereby to the phenomenon of gravitational lensing, in which multiple images of the same distant astronomical object are visible in the sky. Moreover, a black hole is merely an object, with a mass that is so big that photons with the constant speed of light, c_0, cannot escape its gravitational field, but it does not mean that other particles, such as free quarks, cannot escape a black hole. Another effect is the redshift of light due to gravity, when the light moves away from the gravitational field. Since light according to the theory has a mass, $m = \varepsilon_0 \mu_0 hf$, it loses exactly so much of its kinetic energy $mc_0^2/2$, that the wave-particle must offer to escape from the gravitational field, which on the other hand equals its gain of potential energy mgh, where h is the height. As the measurements of time can be related to the frequency of an oscillator, a gravitational field will thus have an influence on the clock-time.

Another question is in connection to the properties of the gravitons. The three other known forces in nature are mediated by virtual particles; electromagnetism by virtual photons, the strong interaction by gluons, and the weak interaction by W and Z bosons, so it is likely that also gravity is mediated by virtual particles. From the equation for the gravitational force $F_G = G(\varepsilon_0 \mu_0 h)^2 f_1 f_2 / r^2$, virtual gravitons is seen to be of the same nature as virtual photons, with the speed equal to the speed of light.

As gravity has the final propagation velocity c_0 relative to the ZPF, masses that are bound together by the gravitational force F_G, and move along the line of force with the common velocity v relative to the zero-point field, will experience an enhanced attractive force equal to $F'_G = Gm_1 m_2 / [r^2(1 - v^2/c_0^2)]$ in the direction of motion, which also can be expressed as $F'_G = G(\varepsilon_0 \mu_0 h)^2 f_1 f_2 / [r^2(1 - v^2/c_0^2)]$. Besides, relativistic gravitational forces also arise when solid bodies move across the line of force that binds the bodies together, where the forces likewise depend on the different distances the forces will have to move from m_1 to m_2 and from m_2 to m_1.

From the final propagation velocity of the electromagnetic and gravitational forces relative to the ZPF, it has been shown that it is possible to derive all the relativistic equations, except for the time dilation, since a physical length contraction cannot be the reason for a time dilation. That is to say, the flow of time is a linear function, so the space is Euclidean.

This can also be seen from that the time is a measure of duration, which is why it can be expressed as $t = x/c$. In connection with a length contraction $x' = x\sqrt{1 - v^2/c_0^2}$ it must therefore take a lesser time t' to move the lesser distance x' with the same velocity v, so t' must be equal to: $t' = t\sqrt{1 - v^2/c_0^2}$, from which it can be seen that the time goes just as fast in the moving system as in the stationary system:

$$c = x'/t' = \frac{x\sqrt{1 - v^2/c_0^2}}{t\sqrt{1 - v^2/c_0^2}} = x/t\,.$$

As the time axis is linear, gravity cannot be a result of the curvature of space-time, which means that the space is Euclidean, and since, for the most part, the cosmological redshift is due to a plasma redshift, which can also be seen from the heating of the Sun's corona, the universes do not expand.

So, the Quantum Ether Theory can explain all the relativistic physical phenomena, such as, for instance, the length contraction, the clock-time dilation, the relativistic mass, the mass-energy equivalence, the gravitational lensing, the gravitational redshift or blueshift of light, the gravitational time dilation and the relativistic gravitational force. It is among other things these properties of time and space that are the basis of the Euclidean Cosmos Theory.

The results of the quantum ether theory are:

- Bodies, which are held together by forces that propagate with a constant velocity in the ZPF, are exposed to a contraction, when they move relative to the field.
- The time is absolute and universal.
- The relativistic mass is equal to $m = m_0(1 - v^2/c_0^2)^{-1/2}$.
- Mass and energy are equivalent quantities, $E = mc_0^2$.
- A black hole in the Euclidean space has a Schwarzschild radius $r = 2GM/c_0^2$.
- The Cosmological Redshift is a combination of the plasma redshift, the Doppler effect, and the gravitational redshift.
- The gravitational redshift: $z \approx GM/(c_0^2 r)$.
- Clocks go slower in a gravitational field than outside, $t' = t(1 - GM/(c_0^2 r))$.
- Due to a length contraction the clocks may be wrong, when they move relative to the ZPF.
- Matter and energy are deflected in a gravitational field.
- A wave-particle has a mass equal to $m = \varepsilon_0 \mu_0 h f$.
- Gravitational lensing is due to the mass of wave-particles: $m = \varepsilon_0 \mu_0 h f$.
- The gravitational force between two wave-particles: $F_G = G(\varepsilon_0 \mu_0 h)^2 f_1 f_2 / r^2$.
- The gravitational force (relativistic): $F'_G = G(\varepsilon_0 \mu_0 h)^2 f_1 f_2 / [r^2(1 - v^2/c_0^2)]$.
- Coulomb's law (relativistic): $F'_C = k_0 q_1 q_2 / r'^2 = k_0 q_1 q_2 / [r^2(1 - v^2/c_0^2)]$.

5.1.2 The Euclidean Cosmos Theory

The Euclidean Cosmos Theory (ECT) explains all the questions in relation to the Cosmos based on classical physics and the quantum theory. The theory deduces the composition of the Cosmos on the basis of the most well-established physical laws and finds that the Cosmos consists of a flat coherent space, which contains the quantum fields plus matter and energy. The constant amount of matter and energy manifests itself as wave-particles, while the quantum fields can be described as virtual particles, which transmit the forces between the various forms of matter. Given that, according to the Quantum Ether Theory (QET), the Cosmos consists of an infinite three-dimensional Euclidean space, where all the matter and energy are collected, it can best be described by a three-dimensional space and a linear time.

When we look at the distribution of matter and energy in the three-dimensional space, we assume that: the Cosmos has existed for an infinite length of time, that no EM interactions travel faster than the speed of light in vacuum and that the amount of matter and energy is constant and quantized - which is why it cannot end up as a singularity. The gravitational force, which is shown to travel with the speed of light, will then produce a mass distribution in the infinite flat space, where matter and energy assemble into barren objects and closed universes, which find themselves in a state of dynamic equilibrium in the Euclidean space, which means that it may happen that barren objects, black holes and universes collide.

Since the Cosmos has always existed, there has been plenty of time for matter and energy to gather into galaxies inside the closed universes. Due to the gravitational force between the galaxies, they have, according to the principle of least effort, organized themselves into large entities, which again have accumulated into the honeycomb structures with their great walls and filaments, which are the largest known structures in the universes.

As matter and energy in each of the galaxies gravitates towards the center of mass, most galaxies have an huge black hole at the center, called an active galactic nucleus, which contains a neutron star at the center. Due to the steady accumulation of matter onto the black hole at the center of a galaxy, the pressure inside the black hole will rise, whereby the neutron star at the center of the black hole gets bigger and denser. At some point the density of the neutrons exceeds the Tolman-Oppenheimer-Volkoff limit, beyond which the neutron star cannot be supported by neutron degeneracy pressure, so the neutron star collapses, during which the extremely high pressure forces the nuclear particles to dissolve into free quarks and gluons, called a quark-gluon plasma. This process releases a vast amount of energy, so if the entire neutron star with a radius of 11.5 km explodes, it would release an energy in the order of 3.5×10^{60} MeV. Such explosions are known as active galactic nuclei, or AGN.

As a black hole is a consequence of Newton's gravitational law, free particles such as free quarks - which are not bound together by forces that propagate with the final speed of light in the ZPF - can move with velocities up to $\sqrt{3}c_0$ in a central gravitational field, and thereby achieve the necessary velocity to leave the black hole. When the quarks and gluons escape from the pressure of the black hole in the form of a jet, they immediately transform into a mixture of nucleons, leptons and electromagnetic radiation. Dependent on the structure of the black hole, its rotation velocity and the magnetic field of the AGN, the eruptions may take many different forms such as quasars, pulsars or magnetars.

The energy productions at the center of the galaxies, where the largest black holes are located, are the most powerful sources of energy in the Cosmos, where the fission of each neutron into two down quarks and one up quark releases an energy equivalent to 928 MeV, while a whole chain reaction in the Sun by comparison only releases 26.73 MeV, [226] Such AGNs produce the largest eruptions and spread the nucleons, electrons and photons into the surroundings, where a part end up in the interstellar medium, where they supply the molecular clouds with new energy, while the rest end up in the intergalactic medium. Such processes, where heavier elements transform into lighter ones, are called regenerative processes.

Since the regenerative processes are far larger in the largest galaxies, there is an upper limit to the size of the galaxies. So even when the galaxies collide at all sorts of different angles, which gives rise to many different types of galaxies, the galaxies will eventually get rid of the superfluous matter due to their enhanced activity. This can also be seen from the tight relation between the size of a black hole and the brightness of a galaxy, which entails that, the greater the black hole becomes, the more material the galaxy spews out. If this connection did not set an upper limit to the size of the galaxies, they would, due to gravity, grew out of all proportions.

In spiral galaxies, such as our own, the mass distribution in the plane perpendicular to the axis of rotation follows the centrifugal principle, which means that the force exerted by rotational motion separates the components in relation to their mass, so the lighter particles, relative to the mean value of the mass of the particles, approaches the center of the galaxy, where they supply the fuel to the luminous bulge mass, while the heavier black holes are flung towards the periphery, where they constitute most of the dark matter, called the halo.

Since the Cosmos has always existed, the dark matter originates from the fact that it only requires about 3.2 $M_\odot$ to create a black hole. In each of the galaxies the black holes make up the structure of the galaxies, and each time a black hole with a neutron star at the center, accumulates so much material that it generates a regenerative process, the generated lighter matter is drawn towards the center of the galaxy. So, it will always be at the center of the galaxies that most of the activity takes place, while the black holes outside the center of a galaxy act as dark matter.

Due to the regenerative processes the universes are in a kind of equilibrium where the matter and energy are distributed over a huge area so the densities of the universes are rather poor and scattered, which can be seen from our own Universe, where the the density is around 10 hydrogen atoms per cubic meter, or approximately 10^{-26} kg/m^3. The intergalactic medium consists of a plasma of electrons, protons and alpha particles, where the distinct separation of ions and electrons produces an electric field, which in turn produces electric currents and magnetic fields. When light traverses the sparse hot plasma, the light becomes red-shifted due to its loss of energy to the plasma, just as electrons lose energy when they traverse an electromagnetic field. [225] This redshift explains the warming of the Sun's corona, the hot intergalactic plasma and is the basis for the Hubble Law, so the redshift is not a consequence of an expansion of our Universe, which on the contrary is almost static.

The gas nebulae, which are interstellar clouds of dust, hydrogen, helium and other ionized gases, stem mainly from the black holes at the center of the galaxies, and are the first step on the road of stars, giants, white dwarfs, supernovae, neutron stars and black holes, where the energy once again ends up at the center of the galaxy. As the CMB consists of the electromagnetic radiation from the regenerative processes in the universe, it reflects the structure of the universe with the great walls and large voids.

Our perception of the size of our Universe stems from the fact that there is a definite limit to what we can observe. The light we receive from the surroundings decreases with the square of the distance, so, at the distance of our observable Universe at 13.8×10^9 ly, we receive so few photons from even the largest and most luminous objects, such as galaxies and quasars, that even the most powerful telescopes cannot distinguish the photons from the background noise.

Since the universes would disappear if they did not hold on to their matter and energy, they must be closed. But it may, of cause, transpire that the universes receive energy from the outside. When a universe has reached its maximum size, which is determined by its density, the excess energy will escape from the closed universe, so it is likely that most of the universes have achieved the same maximum size. So each of the universes has a final size determined by their mass and density distribution.

It is thus possible to provide an estimation of the size and mass of our Universe, based on a density function and the observed mass density of both ordinary matter and dark matter. Since, due to the regenerative processes, the density is rather small, the density distribution of the Sun has been used as a preliminary estimate of the density function of our Universe. Based on these premises, it has been found that the radius of our Universe equals 2.6×10^{27} m, while the mass including dark matter is found to be equal to 2.7×10^{55} kg. For comparison, the radius of our observable Universe equals 4.4×10^{26} m.

Since the universes have some of the same properties as the galaxies, they will probably organize themselves as the galaxies have done. In each of the universes there is a balance between the outwardly directed forces generated by the regenerative processes and the inwardly directed gravitational forces. This equilibrium generates a density distribution of the universes that is comparable to the density distribution of a star. As, due to gravity, the density is greatest at the center, the gravitational force at the periphery becomes so small that the universes cannot hold on to any additional radiation, so the universes have a final extent, which is the same for all the universes that have reached their maximum size. The outwardly directed radiation created by the regenerative processes may be interpreted as an expansion of our universe, while we hardly experience the inward flow of the large structures.

As the universes grow through merger with all the material they are able to attract, until they reach a size where they are no longer able to hold on to matter and energy, the superfluous material will end up in the space between the universes. Consequently, there will be an eternal exchange of matter and energy between the universes. Since the universes, just as the galaxies, get rid of the superfluous material when they reach a certain size, the universes will organize themselves in the same way as the galaxies. So the universes will, depending on their number, most likely organize themselves in honeycomb structures.

We can finally conclude that classical physics together with the quantum theory is able to provide a complete scientific description of the relativistic phenomena and the quantum mechanical nature of the Cosmos. These fundamental theories have thus made it possible to determine the composition of the Cosmos, which in every respect fits with the observations of our Universe. So, the dissertation is not just another theory of the Universe, but a determination of the nature of the Cosmos based on the most solid physics we know, which makes it possible not just to observe, but also to understand, virtually all the cosmological phenomena and observations.

The results of the Euclidean Cosmos Theory:

- The theory explains the structure of the Cosmos.
- The theory explains the superstructures of the universes.
- The theory explains the mass distribution in galaxies.
- The theory explains the seemingly eternal life of the galaxies.
- The theory explains how regenerative explosions are produced.
- The theory provides an explanation of why our Universe is flat.
- The theory explains why our Universe is not expanding.

- The theory solves the horizon problem.
- The theory solves the smoothness problem.
- The theory explains the origin of the cosmic microwave background.
- The theory explains the temperature fluctuations of the CMB.

5.2 Bibliography

Bibliography

[1] Albert Einstein. *Zur Elektrodynamik bewegter Körper.* Annalen der Physik 17: 891, 1905.

[2] Hendrik Antoon Lorentz. *Electromagnetic phenomena in a system moving with any velocity smaller than that of light.* Proceedings of the Royal Netherlands Academy of Arts and Sciences, 6: 809–831, 1904.

[3] Hendrik Casimir. *On the attraction between two perfectly conducting plates.* Proceedings of the Royal Netherlands Academy, 51, 793-795, 1948.

[4] J. J. O'Connor and E. F. Robertson, James Clerk Maxwell. *School of Mathematics and Statistics, 1862.*

[5] Stefan Marinov. *New Measurement of the Earth's Absolute Velocity with the Help of the "Coupled Shutters" Experiment.* Progress in Physics, 2007.

[6] G. Hinshaw et al. *Five-year Wilkinson Microwave Anisotropy Probe observations: data processing, sky maps, and basic results.* The Astrophysical Journal Supplement Series, 180:225–245, doi:10.1088/0067-0049/180/2/225, Feb 2009.

[7] Albert Einstein et al. *The Principle of Relativity.* Dover Publications, New York, 1952.

[8] M. Planck *On the Law of the Energy Distribution in the Normal Spectrum.* Deutsche Physikalische Gesellschaft, Berlin, 1900.

[9] R. Eisberg and R. Resnick. *Quantum Physics of Atoms, Molecules, Solids, Nuclei, and Particles.* (2nd ed.). John Wiley & Sons. pp. 59–60. ISBN 047187373X, 1985.

[10] I. Bernard Cohen and Anne Whitman. *Isaac Newton, The Principia: Mathematical Principles of Natural Philosophy. Preceded by A Guide to Newton's Principia.* University of California Press, 1999.

[11] J. R. Mayer, J. P. Joule, S. Carnot. *The Discovery of the Law of Conservation of Energy.* Isis, Vol. 13, No. 1, Sep., 1929.

[12] Phillippe H. Eberhard and Ronald R. Ross. *Quantum Field Theory Cannot Provide Faster-than-light Communication.* Lawrence Berkeley Laboratory, University of California Berkeley, California 94720.

[13] Bombaci. *The Maximum Mass of a Neutron Star.* Astronomy and Astrophysics. Bibcode:1996A&A...305..871B, 1996.

[14] Henrik Bohr, H. B. Nielsen. *Hadron production from a boiling quark soup: quark model predicting particle ratios in hadronic collisions.* Nuclear Physics B. 128 (2): 275, 1977.

[15] H. D. Politzer. *Asymptotic Freedom: An Approach to Strong Interactions.* Physics Reports 14: 129, 1974.

[16] F. Wilczek, D. J. Gross. *Asymptotically Free Gauge Theories.* Phys. Rev. D 8, 3633, 1973.

[17] John Bally, Bo Reipurth. *The Birth of Stars and Planets.* Cambridge University Press. p. 207. ISBN 978-0-521-80105-8, 2006.

[18] Alan P. Lightman. *Ancient Light: Our Changing View of the Universe.* Harvard University Press, 1 January 1993.

[19] C. L. Bennett et al. *Nine-Year Wilkinson Microwave Anisotropy Probe (WMAP) Observations: Final Maps and Results.* ApJS., 208, 20B, 2013.

[20] F. Mandl, G. Shaw. *Quantum Field Theory.* John Wiley & Sons, Chichester UK, revised edition, 1984/2002.

[21] James Clerk Maxwell. *A Dynamical Theory of the Electromagnetic Field.* Royal Society Transactions, Vol. CLV, 1865.

[22] Paul J. Nahin. *Oliver Heaviside.* University of New Hampshire. The Johns Hopkins University Press, Baltimore and London, 1988.

[23] Peter J. Mohr, Barry N. Taylor, David B. Newell. *CODATA Recommended Values of the Fundamental Physical Constants.* 2008.

[24] John R. Reitz, Frederick J. Milford, Robert W. Christy. *Foundations of Electromagnetic Theory.* Third Edition, Addison-Wesley Publishing Company, 1979.

[25] Diego Miramontes Delgado. *Measuring the Magnetic Permeability Constant mu_0 using a Current Balance.* Physics Department, The College of Wooster, Wooster, Ohio, 44691, USA, June, 2015.

[26] H. Fizeau. *Sur les hypothèses relatives à l'éther lumineux.* Comptes Rendus 33: 349–355, 1851.

[27] J. C. Maxwell. *On Physical Lines of Force.* Philosophical Magazine and Journal of Science, London, Edinburgh and Dublin, 1861.

[28] Coulomb. *Premier mémoire sur l'électricité et le magnétisme.* Histoire de l'Académie Royale des Sciences, pages 569-577, 1785.

[29] C. F. Gauss. *Carl Friedrich Gauss Werke.* Königlichen Gesellschaft der wissenschaften, 1877.

[30] L. H. Ryder. *Quantum Field Theory.* Cambridge University Press, 1985.

[31] Morten Scharff. *Elementær Kvantemekanik.* Akademisk Forlag, 1971.

[32] R. Mirman. *Quantum Mechanics, Quantum Field Theory.* Huntington NY, Nova Science Publishers, 2001.

[33] L. De Broglie. *The reinterpretation of wave mechanics, Foundations of Physics.* 1(1): 5–15, p. 9, 1970.

[34] L. Mandelshtam, I. Tamm. *The uncertainty relation between energy and time in nonrelativistic quantum mechanics.* (Ser. Fiz.) 9: 122–128, Izv. Akad, Nauk, SSSR, 1945.

[35] Richard Feynman. *QED: The Strange Theory of Light and Matter.* Princeton University Press, chapter 1, p. 6, Princeton, 1985.

[36] CODATA Value: Planck constant. *National Institute of Standards and Technology.* U.S. Department of Commerce, 2014.

[37] The International System of Units (SI), 8th edition 2006. *Organisation Intergouvernementale de la Convention du Mètre.* STEDI MEDIA, 1, Boulevard Ney, 75018 Paris, 2006.

[38] Erik Curiel and Peter Bokulich. *Singularities and Black Holes.* Stanford Encyclopedia of Philosophy.

[39] Ture Eriksson, Torbjörn Lagerwall and Olof Backman. *Mekanik. Värmeläre.* Almqvist & Wiksell Förlag AB, Stockholm, 1970.

[40] Arthur H. Compton. *A Quantum Theory of the Scattering of X-Rays by Light Elements.* Physical Review, 21 (5): 483–502, May 1923.

[41] Peter J. Mohr and Barry N. Taylor. *National Institute of Standards and Technology.* U.S. Department of Commerce, 2002.

[42] A. Brynjolfsson. *Redshift of photons penetrating a hot plasma.* Applied Radiation Industries, 7 Bridle Path, Wayland, MA 01778, USA, arXiv:astro-ph/0401420, 2004.

[43] Fraser Cain. *What is Intergalactic Space?*. Universe Today, 2014.

[44] M. Tadokoro. *A Study of the Local Group by Use of the Virial Theorem.* Publications of the Astronomical Society of Japan, 1968.

[45] David T. Chuss. *Cosmic Background Explorer.* NASA Goddard Space Flight Center, 2008.

[46] A. M. Bykov, F. B. S. Paerels, V. Petrosian. *Equilibration Processes in the Warm-Hot Intergalactic Medium.* Space Science Reviews, 134 (1–4): 141–153, arXiv:0801.1008, 2008.

[47] K. J. H. Phillips. *Guide to the Sun.* Cambridge University Press. p. 295. ISBN 978-0-521-39788-9. 1995.

[48] M. J. Aschwanden. *Physics of the Solar Corona. An Introduction. Praxis Publishing.* ISBN 3-540-22321-5, 2004.

[49] A. Piel. *Plasma Physics: An Introduction to Laboratory, Space, and Fusion Plasmas.* Springer. pp. 4–5. ISBN 978-3-642-10491-6, 2010.

[50] W. Heitler. *The Quantum Theory of Radiation.* 3rd ed. Oxford Clarendon Press, 1954.

[51] R. J. Gould. *The cross section for double Compton scattering.* Astrophysical Journal, Part 1 (ISSN 0004-637X), vol. 285, Oct. 1, 1984, p. 275-278.

[52] Hugh D. Young, Roger A. Freedman, Lewis Ford. *Sears and Zemansky's university physics.* Bind 1, Addison-Wesley series in physics, Addison-Wesley, 2006.

[53] A. Riess et al. *Observational Evidence from Supernovae for an Accelerating Universe and a Cosmological Constant.* The Astronomical Journal. 116 (3): 1009–1038. arXiv:astro-ph/9805201. doi:10.1086/300499, 1998.

[54] Tod R. Lauer and Marc Postman. *The motion of the Local Group with respect to the 15,000 kilometer per second Abell cluster inertial frame.* Astrophysical Journal, Part 1 (ISSN 0004-637X), vol. 425, no. 2, p. 418-438, 1994.

[55] A. Brynjolfsson. *Hubble constant from lensing in plasma-redshift cosmology, and intrinsic redshift of quasars.* arXiv:astro-ph/0411666v3, 2004.

[56] Planck Collaboration. *Planck 2015 results. XIII. Cosmological parameters.* arXiv: 1502.01589, 2015.

[57] Itzhak Bars, John Terning. *Extra Dimensions in Space and Time.* Springer, ISBN 978-0-387-77637-8, May 2011.

[58] D. Palmer. *Hydrogen in the Universe.* NASA, Feb., 2008.

[59] Eric J. Simon, Jean L. Dickey, Kelly A. Hogan, Jane B. Reece. *Campbell Biology: Concepts & Connections.* Pearson, ISBN-13: 9780134296012, 2018.

[60] O. Heaviside. *Electromagnetic Theory.* Vol. 1, p. 455-466, Apendix B, 1893.

[61] P. Anninos, D. Hobill, E. Seidel, L. Smarr, W-M. Suen. *The Collision of Two Black Holes.* National Center for Supercomputing Applications, Beckman Institute; Department of Physics and Astronomy, University of Calgary; McDonnell Center for the Space Sciences, Department of Physics, Washington University.

[62] R. A. Matzner, H. E. Seidel, S. L. Shapiro, L. Smarr, W-M. Suen, S. A. Teukolsky, J. Winicour. *Geometry of a Black Hole Collision.* Science 10 November 1995: Vol. 270. no. 5238.

[63] M. Milosavljevic, E. S. Phinney. *The Afterglow of Massive Black Hole Coalescence.* Theoretical Astrophysics, California Institute of Technology, 2008.

[64] C. W. Misner, K. S. Thorne, J. A. Wheeler. *Gravitation.* University of Maryland; California Institute of Technology; Universiy of Texas, Austin; Princeton University, W. H. Freeman, New York, 1973.

[65] Wenwen Zuo, Xue-Bing Wu, Xiaohui Fan, Richard Green, Ran Wang, Fuyan Bian. *Black Hole Mass Estimates and Rapid Growth of Supermassive Black Holes in Luminous $z \sim 3.5$ Quasars.* The Astrophysical Journal. 799 (2): 189, arXiv:1412.2438, 2014.

[66] M. M. Aggarwal et al. *An Experimental Exploration of the QCD Phase Diagram: The Search for the Critical Point and the Onset of De-confinement.* arXiv:1007.2613v1 [nucl-ex], 15 Jul 2010.

[67] *Notes on Dirac's lecture Developments in Atomic Theory.* Le Palais de la Découverte, 6 December 1945.

[68] Richard M. Weiner. *The Mysteries of Fermions.* International Journal of Theoretical Physics. 49 (5): 1174–1180, arXiv:0901.3816, 2010.

[69] Andrew G. Truscott, Kevin E. Strecker, William I. McAlexander, Guthrie Partridge and Randall G. Hulet. *Observation of Fermi Pressure in a Gas of Trapped Atoms.* Science, 2 March 2001.

[70] G. Fontaine, P. Brassard, P. Bergeron. *The Potential of White Dwarf Cosmochronology.* Publications of the Astronomical Society of the Pacific. 113 (782): 409–435. doi:10.1086/319535, 2001.

[71] M. Richmond. *Late stages of evolution for low-mass stars.* Lecture notes, Physics 230. Rochester Institute of Technology, 3 May 2007.

[72] A. S. Eddington. *On the relation between the masses and luminosities of the stars.* Monthly Notices of the Royal Astronomical Society. 84 (5): 308–333. doi:10.1093/mnras/84.5.308, 1924.

[73] J. Johnson. *Extreme Stars: White Dwarfs & Neutron Stars.* Lecture notes, Astronomy 162. Ohio State University, 2007.

[74] R. H. Fowler. *On dense matter.* Monthly Notices of the Royal Astronomical Society. 87 (2): 114–122. doi:10.1093/mnras/87.2.114, 1926.

[75] Paolo A. Mazzali, Friedrich K. Röpke, Stefano Benetti, Wolfgang Hillebrandt. *A Common Explosion Mechanism for Type Ia Supernovae.* Science, 315 (5813): 825–828. arXiv:astro-ph/0702351, 2007.

[76] A. Heger, C. L. Fryer, S. E. Woosley, N. Langer, D. H. Hartmann. *How Massive Single Stars End Their Life.* The Astrophysical Journal. 591: 288–300. arXiv:astro-ph/0212469, 2003.

[77] Norman K. Glendenning. *Compact Stars: Nuclear Physics, Particle Physics and General Relativity.* Springer Science & Business Media. p. 1. ISBN 978-1-4684-0491-3, 2012.

[78] N. Chamel, Paweł Haensel, J. L. Zdunik, A. F. Fantina. *On the Maximum Mass of Neutron Stars.* International Journal of Modern Physics. 1 (28): 1330018. arXiv:1307.3995. doi:10.1142/S021830131330018X, Nov 2013.

[79] M. Coleman Miller and Jon M. Miller. *The Masses and Spins of Neutron Stars and Stellar-Mass Black Holes.* University of Maryland and Michigan, arXiv:1408.4145, Aug 2014.

[80] W. Plessas and L. Mathelitsch. *Lectures on quark matter.* Lecture notes in physics (Vol. 583), Springer, 2002.

[81] F. J. Dyson, A. Lenard. *Stability of Matter I.* J. Math. Phys. 8 (3): 423–434. Bibcode:1967JMP.....8..423D. doi:10.1063/1.1705209, 1967.

[82] F. J. Dyson. *Ground-State Energy of a Finite System of Charged Particles.* J. Math. Phys. 8 (8): 1538–1545. Bibcode:1967JMP.....8.1538D. doi:10.1063/1.1705389, 1967.

[83] H. S. Goldberg, M. D. Scadron. *Physics of Stellar Evolution and Cosmology.* Taylor & Francis. ISBN 0-677-05540-4, 1987.

[84] M. Bachetti, F. A. Harrison, D. J. Walton, B. W. Grefenstette, D. Chakrabarty, F. Fürst, D. Barret, A. Beloborodov, S. E. Boggs, F. E. Christensen, W. W. Craig, A. C. Fabian, C. J. Hailey, A. Hornschemeier, V. Kaspi, S. R. Kulkarni, T. Maccarone, J. M. Miller, V. Rana, D. Stern, S. P. Tendulkar, J. Tomsick, N. A. Webb, W. W. Zhang. *An ultraluminous X-ray source powered by an accreting neutron star.* Nature; 514 (7521): 202 doi:10.1038/nature13791, 2014.

[85] Patrick Betts. *Astrophysics.* PediaPress, 2014.

[86] Archive. *Neutron degeneracy pressure.* Physics Forums. Retrieved on 2011-10-09.

[87] J. L. Zdunik and P. Haensel. *Maximum mass of neutron stars and strange neutron-star cores.* N. Copernicus Astronomical Center, Polish Academy of Sciences, Warszawa, Poland, 2013.

[88] A. Y. Potekhin. *The Physics of Neutron Stars.* arXiv:1102.5735, 2011.

[89] J. E. Staff, R. Ouyed, and P. Jaikumar. *Quark deconfinement in neutron star cores: The effects of spin-down.* Astrophysical Journal. 645: L145-L148. arXiv:astro-ph/0603743, 2006.

[90] Ignazio Bombaci, Domenico Logoteta, Isaac Vidana, and Constanca Providencia *Quark matter nucleation in neutron stars and astrophysical implications.* Dipartimento di Fisica "E. Fermi", Universita di Pisa, Italy. arXiv:1601.04559v1, 2016.

[91] Krishna Rajagopal. *Mapping the QCD Phase Diagram.* Center for Theoretical Physics, MIT, Cambridge, MA 02139, USA, arXiv:hep-ph/9908360v1, 15 Aug 1999.

[92] Shapiro and Teukolsky. *Black Holes, White Dwarfs and Neutron Stars: The Physics of Compact Objects.* Wiley 2008.

[93] Ignazio Bombaci and Bhaskar Datta. *Conversion of Neutron Stars to Strange Stars as the Central Engine of Gamma-Ray Bursts.* The Astrophysical Journal Letters, Volume 530, Number 2, Jan 2000.

[94] Z. Berezhiani, I. Bombaci, A. Drago, F. Frontera, A. Lavagno. *Isospin dependence of quark deconfinement in heavy ion collisions.* Astrophys. Jour. 586, 1250, 2003.

[95] K. Rajagopal. *Quark-Gluon Plasma 2.* Edited by R. Hwa, World Scientific. arXiv:hep-ph/9504310, 1995.

[96] J. Beringer et al. *PDGLive Particle Summary `Quarks (u, d, s, c, b, t, b´, t´, Free).* Particle Data Group, 2013.

[97] E. Laermann. *Thermodynamics using Wilson and Staggered Quarks.* Nucl. Phys. Proc. Suppl. 63 (1998) 114-125, arXiv:hep-lat/9802030, Feb 1998.

[98] S. Gottlieb et al. *Thermodynamics of lattice QCD with two light quarks on a* $16^3 \times 8$ *lattice.* Phys. Rev. D55, 6852, June 1997.

[99] B. Kiziltan. *Reassessing the fundamentals: On the evolution, ages and masses of neutron stars.* Universal-Publishers, ISBN 1-61233-765-1, arXiv: 1102.5094v1, 2011.

[100] Kumiko Koteraa, Elena Amatob, Pasquale Blasi. *The fate of ultrahigh energy nuclei in the immediate environment of young fast-rotating pulsars.* Institut d'Astrophysique de Paris, arXiv:1503.07907v2 [astro-ph.HE] Jul 2015.

[101] Fridolin Weber, Gustavo A. Contrera, Milva G. Orsaria, William Spinella, and Omair Zubairi. *Properties of high-density matter in neutron stars.* doi:10.1142/S0217732314300225, arXiv:1408.0079 [astro-ph.SR], Aug 2014.

[102] Takashi J. Moriya. *Progenitors of recombining supernova remnants.* arXiv:1203.5799 [astro-ph.HE] Mar 2012.

[103] V. S. Beskin. *Radio pulsars.* Uspekhi Fizicheskikh Nauk and Russian Academy of Sciences, Physics-Uspekhi, Volume 42, Number 11, 1999.

[104] M. Servillat et all. *Neutron star atmosphere composition: the quiescent, low-mass X-ray binary in the globular cluster M28.* arXiv:1203.5807, Mar 2012.

[105] A. Reisenegger. *Origin and Evolution of Neutron Star Magnetic Fields.* Departament of Astronomy, Santiago, Chile, 2003.

[106] José A. Pons, Daniele Viganò, Nanda Rea. *Too much "pasta" for pulsars to spin down.* Nature Physics. 9 (7): 431–434. arXiv:1304.6546, 2013.

[107] M. Coleman Miller. *Introduction to neutron stars.* Professor of Astronomy, University of Maryland, 2007.

[108] F. Douchin, P. Haensel. *A unified equation of state of dense matter and neutron star structure.* Department of Physics, University of Illinois at Urbana-Champaign, USA; CRAL, ENS de Lyon, France; CAMK, Warsaw, Poland, arXiv:astro-ph/0111092, Nov 2001.

[109] Feryal Ozel, Paulo Freire. *Masses, Radii, and the Equation of State of Neutron Stars.* Annu. Rev. Astron. Astrophys. 54 (1): 401–440. arXiv:1603.02698. doi:10.1146/annurev-astro-081915-023322, 2016.

[110] Simon F. Green, Mark H. Jones, S. Jocelyn Burnell. *An Introduction to the Sun and Stars.* Cambridge University Press. ISBN 978-0-521-54622-5, 2004.

[111] Corvin Zahn. *Tempolimit Lichtgeschwindigkeit.* 2009.

[112] Datos Freak. *Peligroso lugar para jugar tenis.* Jun 2016.

[113] B. Povh, K. Rith, C. Scholz, F. Zetsche. *Particles and Nuclei: An Introduction to the Physical Concepts.* Berlin: Springer-Verlag. p. 73. ISBN 978-3-540-43823-6, 2002.

[114] P. J. Mohr, B. N. Taylor, and D. B. Newell. *The 2014 CODATA Recommended Values of the Fundamental Physical Constants.* The database was developed by J. Baker, M. Douma, and S. Kotochigova. National Institute of Standards and Technology, Gaithersburg, Maryland 20899, 2014.

[115] J. M. Lattimer and M. Prakash. *The Physics of Neutron Stars.* Department of Physics and Astronomy State University of New York at Stony Brook, USA, arXiv:astro-ph/0405262, May 2004.

[116] The CMS Collaboration. *Search for quark compositeness in dijet angular distributions from pp collisions at sqrt(s) = 7 TeV.* Journal of High Energy Physics. arXiv:1202.5535, 2012.

[117] W. M. Yao et al. *Review of Particle Physics.* Journal of Physics G. 33:1. Particle Data Group, arXiv:astro-ph/0601168, 2006.

[118] Adrian Cho. *Mass of the Common Quark Finally Nailed Down.* Science Magazine, Apr 2010.

[119] Alak Ray. *Massive stars as thermonuclear reactors and their explosions following core collapse.* Tata Institute of Fundamental Research, Mumbai 400 005, India, arXiv:0907.5407v1 [astro-ph.SR], Jul 2009.

[120] IAU Division I Working Group. *Numerical Standards for Fundamental Astronomy.* Astronomical Constants: Current Best Estimates (CBEs), 2012.

[121] S. Gandolfi, J. Carlson, and Sanjay Reddy. *The maximum mass and radius of neutron stars and the nuclear symmetry energy.* Theoretical Division, Los Alamos National Laboratory, and Institute for Nuclear Theory, University of Washington, USA, arXiv:1101.1921v2 [nucl-th], Mar 2012.

[122] V. B. Bhatia. *Astronomy and Astrophysics with Elements of Cosmology.* Department of Physics and Astrophysics, University of Delhi India, ISBN 1-84265-021-1, 2001.

[123] Llanes-Estrada, J. Felipe, Moreno Navarro, and Gaspar. *Cubic neutrons.* arXiv:1108.1859v1, 2011.

[124] Simone Conradi. *Properties of QCD at finite temperature and density.* Dottorato in Fisica XX ciclo Universita di Genova, 2007.

[125] K. Wu, D. P. Menezes, D. B. Melrose, and C. Providencia. *QCD phase transition in neutron stars and gamma-ray bursts.* UCL, University College London, 2006.

[126] J. M. Lattimer, M. Prakash. *Neutron Star Structure and the Equation of State.* Astrophys.J.550:426,2001, ,arXiv:astro-ph/0002232, 2001.

[127] Walter Lewin, Michiel van der Klis. *Compact Stellar X-ray Sources.* Cambridge University Press, ISBN 978-0-521-82659-4, April 2006.

[128] Geert Jan Besjes. *Pushing Susy's Boundaries - Searches and prospects for strongly-produced supersymmetry at the LHC with the ATLAS detector.* Radboud University Nijmegen, 2015.

[129] H. Zheng, J. Sahagun, A. Bonasera. *Neutron stars and supernova explosions in the framework of Landau's theory.* arXiv:1411.3030 [nucl-th], Nov 2014.

[130] Mary K. Gaillard, Paul D. Grannis, and Frank J. Sciulli. *The Standard Model of Particle Physics.* arXiv:hep-ph/9812285, Dec 1998.

[131] E. Cartlidge. *Quarks break free at two trillion degrees.* Physics World, Jun 2011.

[132] V. Barger, R. Phillips. *Collider Physics.* Addison–Wesley, ISBN 0-201-14945-1, 1997.

[133] Belle Dumé. *Quark-gluon plasma goes liquid.* physicsworld.com, Mar 2016.

[134] Samuel Bodman, Raymond L. Orbach, Sam Aronson. *RHIC Scientists Serve Up "Perfect" Liquid.* Brookhaven National Laboratory, Apr 2005.

[135] A. Smilga. *Lectures on quantum chromodynamics.* World Scientific. ISBN 978-981-02-4331-9, 2001.

[136] The Fermi-LAT collaboration: M. Ackermann et al. *Detection of the Characteristic Pion-decay Signature in Supernova Remnants.* arXiv:1302.3307 [astro-ph.HE], Feb 2013.

[137] Chun Shen. *Electromagnetic Radiation from QCD Matter: Theory Overview.* Department of Physics, McGill University, 3600 University Street, Montreal, QC, H3A 2T8, Canada, arXiv:1601.02563v1 [nucl-th], Jan 2016.

[138] E. V. Shuryak. *Quark-gluon plasma and hadronic production of leptons, photons and psions.* Institute of Nuclear Physics, Novosibirsk, USSR, doi.org/10.1016/0370-2693(78)90370-2, Mar 1978.

[139] C. Seife. *High-Energy Physics: Cern Stakes Claim on New State of Matter.* Science Magazine, pages 949 - 951, Volume 287, Number 5455 February 11, 2000.

[140] M. Rees. *Exploring Our Universe and Others.* Scientific American, December 1999.

[141] Nadine Haering, Hans-Walter Rix. *On the Black Hole Mass - Bulge Mass Relation.* Astrophys.J.604:L89-L92,2004, doi:10.1086/383567, arXiv:astro-ph/0402376, Feb 2004.

[142] NASA caption. *Correlation between black hole Mass and bulge mass/brightness.* Id:opa0022b, Artwork, Jun 2000.

[143] J. S. Kaastra and H. G. van Bueren. *Mass-to-energy Relations for Galaxies and Clusters of Galaxies.* Sterrekundig Instituut, Utrecht, The Netherlands, Bibcode: 1981A&A....99....7K, Mar 1981.

[144] A. Marconi, L. K. Hunt. *The Relation between Black Hole Mass, Bulge Mass, and Near-Infrared Luminosity.* The Astrophysical Journal. 589 (1): L21-L24, arXiv:astro-ph/0304274, doi:10.1086/375804, 2003.

[145] NASA. *Most Luminous Galaxy Is Ripping Itself Apart.* Jet Propulsion Laboratory, California Institute of Technology, Jan 2016.

[146] Nick Z. Scoville. *Evolution of star formation and gas.* California Institute of Technology, USA, arXiv:1210.6990v1, Oct 2012.

[147] Eric Herbst. *Chemistry in The Interstellar Medium.* Annual Review of Physical Chemistry. doi:10.1146/annurev.pc.46.100195.000331. Oct 2014.

[148] Kevin Schawinski et al. *Supernova Shock Breakout from a Red Supergiant.* Science. 321 (5886): 223–226. arXiv:0803.3596, 2008.

[149] F. X. Timmes, S. E. Woosley, Thomas A. Weaver. *Galactic chemical evolution: Hydrogen through zinc.* Astrophysical Journal Supplement Series. 98: 617. arXiv:astro-ph/9411003, 1995.

[150] Doug C. B. Whittet. *Dust in the Galactic Environment.* CRC Press. pp. 45–46. ISBN 0-7503-0624-6, 2003.

[151] Luiz C. Jafelice, Reuven Opher. *The origin of intergalactic magnetic fields due to extragalactic jets.* Monthly Notices of the Royal Astronomical Society, Royal Astronomical Society, 257 (1): 135–151, doi:10.1093/mnras/257.1.135, Jul 1992.

[152] W. James Wadsley et al. *The Universe in Hot Gas.* Astronomy Picture of the Day, NASA, archived from the original on June 9, 2009, retrieved 2009-06-19, Aug 2002.

[153] T. Fang et al. *Confirmation of X-Ray Absorption by Warm-Hot Intergalactic Medium in the Sculptor Wall.* The Astrophysical Journal, 714 (2): 1715, arXiv:1001.3692, doi:10.1088/0004-637X/714/2/1715, 2010.

[154] Anjali Gupta, M. Galeazzi, E. Ursino. *Detection and Characterization of the Warm-Hot Intergalactic Medium.* Bulletin of the American Astronomical Society, 41: 908, Bibcode:2010AAS...21631808G, May 2010.

[155] B. P. Wakker, B. D. Savage. *The Relationship Between Intergalactic H I/O VI and Nearby (z<0.017) Galaxies.* The Astrophysical Journal Supplement Series, 182: 378, arXiv:0903.2259, doi:10.1088/0067-0049/182/1/378, 2009.

[156] B. F. Mathiesen, A. E. Evrard. *Four Measures of the Intracluster Medium Temperature and Their Relation to a Cluster's Dynamical State.* The Astrophysical Journal, 546: 100, arXiv:astro-ph/0004309, doi:10.1086/318249, 2001.

[157] R. Antonucci. *Unified Models for Active Galactic Nuclei and Quasars.* Annual Review of Astronomy and Astrophysics. 31 (1): 473-521. doi:10.1146/annurev.aa.31.090193.002353, 1993.

[158] R. Schödel et al. *A star in a 15.2-year orbit around the super-massive black hole at the center of the Milky Way.* Nature. 419 (6908): 694–696. arXiv:astro-ph/0210426, doi:10.1038/nature01121, 2002.

[159] Randall L. Cooper, Ramesh Narayan. *Theoretical Models of Superbursts on Accreting Neutron Stars.* arXiv:astro-ph/0410462, Oct 2004.

[160] J. Krebs, Wolfgang Hillebrandt. *The interaction of supernova shockfronts and nearby interstellar clouds.* Astronomy and Astrophysics. 128: 411. Bibcode:1983A&A...128..411K. 1983.

[161] Scott A. Hughes. *Trust but verify: The case for astrophysical black holes.* arXiv:hep-ph/0511217, 2005.

[162] Donald D. Clayton. *Principles of Stellar Evolution and Nucleosynthesis.* Mc-Graw Hill, New York, 1968.

[163] A. Parikh, J. Jose, G. Sala, C. Iliadis. *Nucleosynthesis in Type I X-ray Bursts.* arXiv:1211.5900, Nov 2012.

[164] Rebecca G. Martin, Chris Nixon, Philip J. Armitage, Stephen H. Lubow, Daniel J. Price. *Giant Outbursts in Be/X-ray Binaries.* arXiv:1407.5676, Jul 2014.

[165] Jean in 't Zand. *Understanding superbursts.* arXiv:1702.04899, Feb 2017.

[166] HESS collaboration. *Acceleration of petaelectronvolt protons in the Galactic Centre.* Nature. 531: 476–479, Natur.531..476H. PMID 26982725, arXiv:1603.07730, doi:10.1038/nature17147, 2016.

[167] G. Cassiday, H. Bergeson, G. Loh, and P. Sokolsky. *The Fly's Eye (1981-1993).* A "New" Experiment, Utah, 1981.

[168] Natalie Wolchover. *The Particle That Broke a Cosmic Speed Limit.* Quanta Magazine, May 2015.

[169] Demosthenes Kazanas. *Toward a Unified AGN Structure". Astronomical Review.* arXiv:1206.5022, doi:10.1080/21672857.2012.11519707, 2012.

[170] K. Wu, D. P. Menezes, D. B. Melrose, and C. Providencia. *QCD phase transition in neutron stars and gamma-ray bursts.* University College London, 2006.

[171] S. E. Woosley. *Gamma-ray bursts from stellar mass accretion disks around black holes.* Astrophysical Journal, Part 1, vol. 405, no. 1, p. 273-277, Mar 1993.

[172] Nancy Atkinson. *New Kind of Gamma Ray Burst is Ultra Long-Lasting.* Universetoday.com, May 2015.

[173] B. Gendre et al. *The Ultra-Long Gamma-Ray Burst 111209A: The Collapse of a Blue Supergiant?* The Astrophysical Journal, Volume 766, Number 1, arXiv:1212.2392, Mar 2013.

[174] G. Vedrenne, J. L. Atteia. *Gamma-Ray Bursts: The brightest explosions in the universe.* Springer, 2009.

[175] P. Podsiadlowski. *Supernovae and Gamma-Ray Bursts.* Springer, 2013.

[176] B. F. Schutz. *Gravity from the ground up.* Cambridge University Press. pp. 98–99. ISBN 978-0-521-45506-0, 2003.

[177] NASA. *What is the Universe Made Of'.* National Aeronautics and Space Administration, Jan 2013.

[178] Luca Casagrande, Don A .VandenBerg. *Synthetic stellar photometry - General considerations and new transformations for broad-band systems.* Monthly Notices of the Royal Astronomical Society, 444, Oxford University Press, pp. 392–419, arXiv:1407.6095, doi:10.1093/mnras/stu1476, 2014.

[179] Jeanne Hopkins. *Glossary of Astronomy and Astrophysics.* The University of Chicago, ISBN 0-226-35171-8, 1980.

[180] P. A. Oesch et all. *A Remarkably Luminous Galaxy at $z = 11.1$ Measured with Hubble Space Telescope Grism Spectroscopy.* The Astrophysical Journal. 819 (2): 129. arXiv:1603.00461, 2016.

[181] Adi Zitrin et all. *Lyman-alpha Emission from a Luminous $z = 8.68$ Galaxy: Implications for Galaxies as Tracers of Cosmic Reionization.* The Astrophysical Journal. 810: L12. arXiv:1507.02679, 2015

[182] NASA. *New Gamma-Ray Burst Smashes Cosmic Distance Record.* Apr 2009.

[183] N. R. Tanvir et al. *A gamma-ray burst at a redshift of z 8.2.* Nature. 461 (7268): 1254. Bibcode:2009Natur.461.1254T, 2009.

[184] Philip Armitage, Stephan Kuvalis. *The Galaxy luminosity function.* JILA, University of Colorado at Boulder, Spring 2004.

[185] David R. Williams. *Sun Fact Sheet - Sun/Earth Comparison.* National Aeronautics and Space Administration, Apr 2014.

[186] G. Bacon. *Most distant galaxy: Hubble breaks cosmic distance record.* Astronomy, Dec 2017.

[187] V. Springel. *Dark Energy Found Stifling Growth in Universe.* Max Planck Institute for Extraterrestrial Physics.

[188] Shiv S. Kumar. *Study of Degeneracy in Very Light Stars.* Astronomical Journal. 67: 579. doi:10.1086/108658, 1962.

[189] Margaret J. Geller, John P. Huchra. *Mapping the Universe.* Science, Vol. 246. no. 4932, pp. 897 - 903, 17 November 1989.

[190] I. Horvath, J. Hakkila, Z. Bagoly. *The largest structure of the Universe, defined by Gamma-Ray Bursts.* arXiv:1311.1104 [astro-ph.CO], Nov 2013.

[191] V. Springel, C. S. Frenk, S. D. M. White. *The large-scale structure of the Universe.* arXiv:astro-ph/0604561v1, Apr 2006.

[192] Erich Regener, Georg Pfotzer. *Vertical Intensity of Cosmic Rays by Threefold Coincidences in the Stratosphere.* Nature. 136 (3444): 718–719. doi:10.1038/136718a0, Nov 1935.

[193] D. J. Fixsen. *The Temperature of the Cosmic Microwave Background.* University of Maryland, Goddard Space Flight Center, MD, USA, The Astrophysical Journal, 707:916–920, Dec 2009.

[194] Richard Fitzpatrick. *Perihelion Precession of Mercury.* Institute for Fusion Studies, The University of Texas at Austin, 2011.

[195] Bernard Schutz. *A First Course in General Relativity.* Cambridge University Press, org/9780521887052, 2009.

[196] Howard E. Bond et al. *HD 140283: A Star In The Solar Neighborhood That Formed Shortly After The Big Bang.* The Astrophysical Journal, 765, L12, 2013.

[197] Sky-map.org. *HD 140283.* Retrieved February 23, 2013.

[198] L. Casagrande, I. Ramírez, J. Meléndez, M. Bessell, and M. Asplund. *An absolutely calibrated Teff scale from the infrared flux method.* A&A, 512, A54, 2010.

[199] Gary J. Ferland et al. *High Metal Enrichments in Luminous Quasars.* The Astrophysical Journal, April 20, 1996.

[200] T. M. Brown, J. Tumlinson, M. Geha, et al. *The Primeval Populations of the Ultra-Faint Dwarf Galaxies.* ApJ, 753, L21, 2012.

[201] Duccio Macchetto. *Black Holes, Quasars and Active Galaxies.* ESA, Hubble News, 2011.

[202] Robert Irion. *A Quasar in Every Galaxy?* Sky and Telescope, New Track Media, 2014.

[203] Claus Grupen and Glen Cowan. *Astroparticle physics.* Springer. pp. 11–12. ISBN 3-540-25312-2, 2005.

[204] D. E. Thomsen. *End of the World: You Won't Feel a Thing.* Science News, 131 (25): 391, 1987.

[205] D. Savage, T. Jones, and R. Villard. *Hubble Surveys the "Homes" of Quasars.* NASA Release: 96-244, 1996.

[206] W. Wamsteker, R. Albrecht, H. J. Haubold. *Developing Basic Space Science World-Wide.* Springer, XVI, 504 p., 2004.

[207] A. Myers et al. *Clustering Analyses of 300,000 Photometrically Classified Quasars. I. Luminosity and Redshift Evolution in Quasar Bias.* The Astrophysical Journal, Volume 658, Number 1, 2007.

[208] Robert J. A. Lambourne. *Relativity, Gravitation and Cosmology.* Cambridge University Press. p. 222. ISBN 0521131383, Cambridge 2012.

[209] Myungshin et al. *Discovery of 40 Bright Quasars Near the Galactic Plane.* Astrophys.J. 664 (2007) 64-70, Mar 2008.

[210] John Dubinski. *The great Milky-Way – Andromeda collision.* Sky and Telescope, 2006.

[211] Y. Juarez et al. *The metallicity of the most distant quasars.* arXiv:0901.0974v2 [astro-ph.CO] 19 Oct 2009.

[212] Dietrich et al. *Quasar Elemental Abundances at High Redshifts.* The Astrophysical Journal, Volume 589, Number 2, 2003.

[213] Forbes and Ponman. *On the relationship between age and dynamics in elliptical galaxies.* Oxford Journals, Science & Mathematics, MNRAS, Volume 309, Issue 3, 1999.

[214] Trager et al. *The Stellar Population Histories of Local Early-Type Galaxies. I. Population Parameters.* The Astrophysical Journal, Volume 119, Number 4, 2000.

[215] A. Bosma. *The Distribution and Kinematics of Neutral Hydrogen in Spiral Galaxies of Various Morphological Types.* Rijksuniversiteit Groningen, PhD, 1978.

[216] Sukanya Chakrabarti. *A New Probe of the Distribution of Dark Matter in Galaxies.* ApJ, 771, 98C, 2013.

[217] Battaglia et al. *The radial velocity dispersion profile of the Galactic halo: constraining the density profile of the dark halo of the Milky Way.* arXiv:astro-ph/0506102, 5 Jun 2005.

[218] Chartas et al. *CHANDRA Detects Relativistic Broad Absorption Lines from AMP 08279+5255.* The Astrophysical Journal, Volume 579, number 1, 2002.

[219] Salucci et al. *Mass function of dormant black holes and the evolution of active galactic nuclei.* Monthly Notices of the Royal Astronomical Society, Oxford Journals, 1998.

[220] Kormendy and Richstone. *Inward Bound - The Search for Supermassive Black Holes in Galactic Nuclei.* Annual Review of Astronomy and Astrophysics, Vol. 33: 581-624, 1995.

[221] Granato et al. *A Physical Model for the Coevolution of QSOs and Their Spheroidal Hosts.* arXiv:astro-ph/0307202v2 16 Sep 2003.

[222] Fan et al. *A Survey of $z > 5.7$ Quasars in the Sloan Digital Sky Survey. II. Discovery of Three Additional Quasars at $z > 6$.* The Astronomical Journal 125, number 4, 2003.

[223] Leo Blitz. *Evolution of the Interstellar Medium.* Astronomical Society of the Pacific, 1906.

[224] Jacob Aron. *Planck shows almost perfect cosmos – plus axis of evil.* New Scientist, 21 March 2013.

[225] S. V. Bulanov, T. Zh. Esirkepov, S. S. Bulanov, J. K. Koga, K. Kondo, and M. Kando. *Radiation Dominated Electromagnetic Shield.* arXiv:1705.00829. [physics.plasm-ph], May 2017.

[226] V. Castellani, S. Degl'Innocenti, G. Fiorentini, M. Lissia, B. Ricci. *Solar neutrinos: beyond standard solar models.* Phys.Rept.281:309-398. arXiv:astro-ph/9606180, doi:10.1016/S0370-1573(96)00032-4, 1997.

Printed in Great Britain
by Amazon